Concrete Craftsman Series

Placing and Finishing Decorative Concrete Flatwork

Concrete Craftsman Series

Placing and Finishing Decorative Concrete Flatwork

Reported by ACI Committee E703

Michael G. Hernandez
Chair

William D. Palmer
Secretary

Scott Anderson
Daniel P. Dorfmueller
James J. Ernzen
Reynold Franklin
Leonard Gagliardi
Beverly Garnant
William R. Nash
Michael V. Pedraza
Katherine M. Pfleger
William R. Phillips
Christian Plue
Thomas Roth
Thomas G. Tyler

Additional recognition to Lance Boyer, who developed a complete initial draft of this document.

CCS-5(16)

American Concrete Institute
Always advancing

Concrete Craftsman Series:
CCS-5 Placing and Finishing Decorative Concrete Flatwork
ISBN 978-1-942727-59-0

Managing Editor: Michael Tholen, Ph.D., P.E.
Art Program: Claire Hiltz
Professional Development Specialist: Tiffany Vojnovski
Technical Editors: Emily Bush, Cherrie Fergusson
Manager, Publishing Services: Barry M. Bergin
Production Editors: Carl Bischof, Kelli Slayden, Kaitlyn Hinman, Tiesha Elam
Page Design & Composition: Ryan Jay
Manufacturing: Marie Fuller

First Printing: December 2015
Printed in Ann Arbor, Michigan.

American Concrete Institute
38800 Country Club Drive
Farmington Hills, MI 48331
USA
www.concrete.org
+1.248.848.3700

CONTENTS

CHAPTER 5: BASIC DECORATIVE CONCRETE COMPONENTS

CHAPTER 6: SURFACE RETARDERS

CHAPTER 7: STAMPED CONCRETE

CHAPTER 9: DECORATIVE TREATMENT OF STAIRS

INTRODUCTION

Concrete contractors place and finish concrete for a variety of applications. Proper placing and finishing of decorative concrete requires specialized skills using decorative concrete products and techniques not required for most concrete placements. These skills are in addition to the proper placing, finishing, and curing skills needed for routine concrete placements that are covered by existing flatwork finisher resources and programs.

The decorative concrete industry is growing fast. As a result, there is a need for education and information from credible sources. The standards of quality for this growing industry must be maintained and increased. This document was produced with the intent of raising the quality of education for the decorative concrete industry. Quality of construction will progress by continually improving the level of craftsmanship.

Many of the skills associated with decorative concrete flatwork are not presently covered in existing concrete flatwork resources. This document supplements existing resources by providing knowledge of the materials, equipment, and techniques required to successfully install decorative concrete flatwork. Knowledge of traditional concrete flatwork techniques is recommended for full understanding of this document. Decorative concrete topics covered herein include:

- Mixture design for decorative concrete mixtures;
- The use of integral color and dry shake color hardeners;
- The identification and uses of decorative finishing tools and equipment (such as stamping mats and stencils);
- The use of surface retarders and set-retarding admixtures to produce exposed aggregate surfaces;
- Stamping and texturing concrete;
- Seeding fresh concrete surfaces with decorative aggregate;
- Decorative treatments for stair treads and risers; and
- Cleaning and sealing methods for decorative concrete.

This document also covers topics that are universal to concrete flatwork construction:

- Jobsite and personal safety;
- Ensuring the design adheres to Americans with Disabilities Act (ADA) guidelines;
- Preplacement site preparation; and

- Identification and uses of concrete finishing tools.

Additionally, two chapters of texture samples and project examples are included. Architectural cast-in-place walls, chemical stains, decorative overlays, decorative scoring, and other topical finishes are not included in this document.

CHAPTER 1: SAFETY

A well-conceived and well-executed safety program is necessary to keep everyone at the jobsite aware of and protected against possible hazards. Failure to follow safety regulations and to use personal protective equipment (PPE) can lead to injuries. In addition to concern for the injured person, accidents can disrupt the flow and quality of work, further impacting the project. Every contracting company should have a written safety policy that sets out clear lines of authority for training and retraining personnel in safety regulations and procedures related to their construction specialty.

While many safety rules are the same for all types of concrete work, decorative concrete work presents special situations. In addition to the minimum requirements of the Occupational Safety and Health Administration (OSHA) Standards for the Construction Industry 29 CFR 1926,[1] these basic rules should always be followed:

- Minimum safety equipment (Fig. 1.1) for all concrete work includes hard hats, safety glasses, safety vests, full-length pants, long-sleeved shirts, sturdy leather footwear (boots or shoes, no sandals), and ear protection. Additional PPE is necessary for specific tasks and activities.
- When working with wet concrete, impervious gloves, impervious boots, and safety glasses should be worn. Fresh concrete can cause chemical burns and eye injuries.
- Hard hats should be worn when any overhead danger is present.
- Safety glasses and face shields should be worn when operating concrete cutting or chipping tools and equipment such as grinders, saws, and chipping hammers.
- Approved respirators should be worn when handling dust-, mist-, or fume-producing products or performing dust-, mist-, or fume-producing operations such as cutting, grinding, or buffing concrete or applying coatings, sprays, colorants, or similar products that may be toxic.
- Hearing protection should be worn when operations or equipment create noise levels that prevent normal conversation at a distance of 3 ft (1 m).
- Portable electric tools should be in good condition and equipped with the appropriate guards. In addition, portable electric tools should be of the double insulated type or protected by a ground fault circuit interrupter. Ground prongs should not be removed.

Fig. 1.1—Hard hat, eye protection, safety vest, long-sleeved shirt, long pants, boots, and ear protection are minimum safety equipment for decorative concrete work. Dust masks should be worn when necessary.

- Extension cords should be at least 14-gauge, approved for hard or extra hard service, and in good condition.
- Extension cords should be kept out of walkways and protected from sharp objects and vehicular traffic.
- When handling wet concrete, an adequate supply of potable water should be available for hand washing and flushing of the eyes.
- Any exposed reinforcing bar that could create an impalement hazard should be protected by approved caps.
- Material Safety Data Sheets for all hazardous materials should be onsite and available for reference.
- Containers of hazardous materials should be labeled in accordance with the OSHA Hazard Communication Standard.[2]
- Solvent-based sealers are flammable and should be kept away from all sources of ignition. When used indoors, the area must be well ventilated.
- Reduce the risk of trip and fall accidents around formwork and exposed reinforcement by barricading the work area. Keep the jobsite clean — even if it's not your job.
- Bend with your knees and lift with your legs, not your back. Lift only what you can comfortably handle.
- Keep your fingers away from the metal joints of a concrete truck chute. Should a finger be caught in the gap of the joint as the chute is dropped from its stored position, it can slice through a finger like scissors through cloth.

Helpful training information is available online through the OSHA Training Institute[3] and the OSHA 2254 document.[4]

REFERENCES

1. Occupational Safety and Health Administration, "Part 1926 – Safety and Health Regulations for Construction," United States Department of Labor, Washington, DC.

2. Occupational Safety and Health Administration, "Part 1910.1200 – Hazard Communication Standard," United States Department of Labor, Washington, DC.

3. Occupational Safety and Health Administration, "OSHA Training Institute," United States Department of Labor, Washington, DC. http://www.osha.gov/dte/oti/index.html. (last accessed 11/22/2015)

4. Occupational Safety and Health Administration, "Training Requirements in OSHA Standards and Training Guidelines," OSHA 2254, United States Department of Labor, Washington, DC, 1998, http://www.osha.gov/Publications/osha2254.pdf. (last accessed 11/22/2015)

CHAPTER 2: LAYOUT CONSIDERATIONS

A contractor's work must comply with Americans with Disabilities Act (ADA) Guidelines.[1] Failure to comply with ADA Guidelines could result in work being rejected or subjected to a lawsuit. Always review plans and as-built conditions to determine if the work to be installed will comply with ADA. Contractors must know the basics of ADA Guidelines as a minimum to avoid problems. An up-to-date ADA reference guide should be purchased and studied to avoid such problems The following is only a brief overview of ADA Guidelines and should not be considered a complete guide.

ADA Guidelines

Purpose of ADA — ADA Guidelines are required by the U.S. Federal Government and set requirements for accessibility by individuals with disabilities to places of public accommodations and commercial facilities. These guidelines are to be applied during the design, construction, and alteration of such buildings and facilities to the extent required by regulations issued by Federal agencies, including the Department of Justice, under the Americans with Disabilities Act of 1990.

Terms used in ADA

ADA Guidelines use the following terms pertaining to exterior concrete work:

Cross Slope — The slope that is perpendicular to the direction of travel.

Curb Ramp — A short ramp cutting through a curb or built up to it.

Detectable Warning — A standardized surface feature built in or applied to walking surfaces or other elements to warn of hazards on a circulation path.

Ramp — A walking surface that has a running slope steeper than 1:20.

Running Slope — The slope that is parallel to the direction of travel.

Tactile — An object that can be perceived using the sense of touch.

Vehicular Way — A route intended for vehicular traffic, such as in a street, driveway, or parking facility.

Walk — An exterior prepared surface for pedestrian use, including pedestrian areas such as plazas and courts.

Requirements for Flatwork

Accessible Route

At least one accessible route within the boundary of the site is required from public transportation stops, accessible parking, accessible passenger loading zones, and public streets or sidewalks to the accessible building entrance that they serve. An accessible route is a continuous, unobstructed path connecting all accessible elements and spaces of a building or facility. Interior accessible routes may include corridors, floors, ramps, elevators, lifts, and clear floor space at fixtures. Exterior accessible routes may include parking access aisles, curb ramps, crosswalks at vehicular ways, walks, ramps, and lifts. To the maximum extent feasible, the accessible route is required to coincide with the route for the general public.

ADA requirements for accessible routes include:

- A minimum clear width for an accessible route of 36 in. (915 mm), with some exceptions;
- A surface for an accessible route that is stable, firm, and slip-resistant. Changes in elevation up to ¼ in. (6 mm) may be vertical and without edge treatment (Fig. 2.1). Changes in elevation between ¼ and ½ in. (6 and 13 mm) are to be beveled with a slope no greater than 1:2 (Fig. 2.2). Changes in elevation greater than ½ in.(13 mm) are required to meet the requirements for ramps; and
- An accessible route with a running slope greater than 1:20 is required to comply with the requirements for ramps.

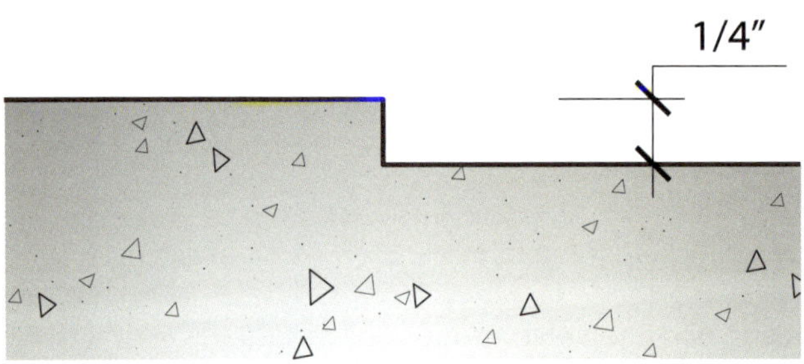

Fig. 2.1—A maximum vertical elevation change of ¼ in. (6 mm) is allowed between accessible route surfaces. (Note: 1 in. = 25.4 mm)

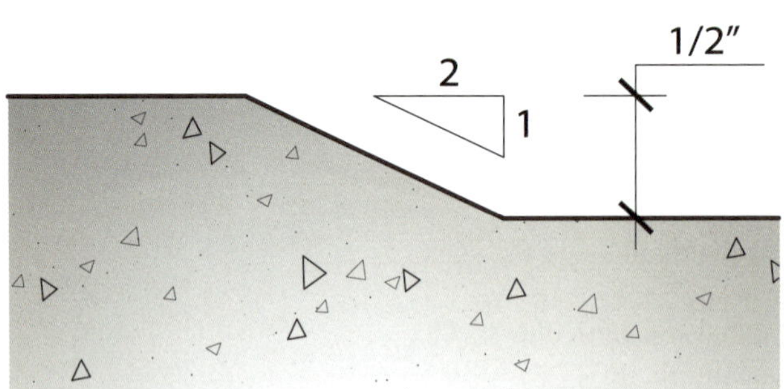

Fig. 2.2—A maximum vertical elevation change of ½ in. (13 mm) is allowed between accessible route surfaces if beveled at a 1:2 slope. (Note: 1 in. = 25.4 mm)

Curb Ramps

Curb ramps are required to be provided wherever an accessible route crosses a curb. ADA requirements for curb ramps include:

- Slopes of curb ramps are required to be at the least possible slope. The maximum slope of a ramp in new construction is 1:12;

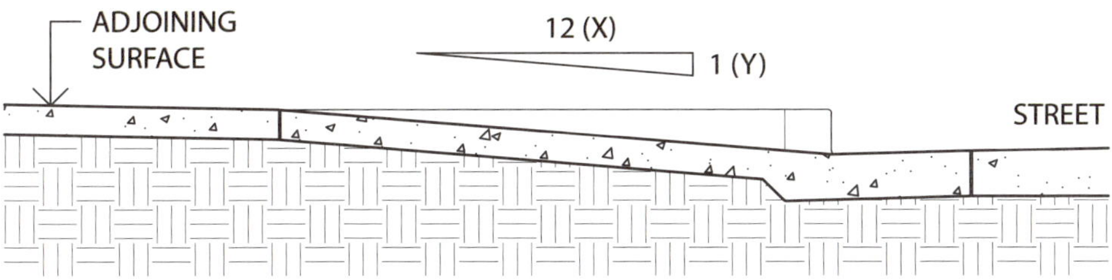

Fig. 2.3—Slope of curb ramp. Slope = Y:X, where X is a level plane.

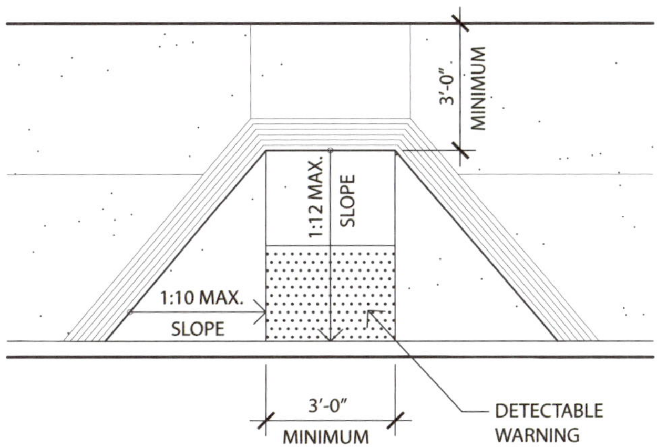

Fig. 2.4—Curb cut ramp (Note: 1 ft = 0.305 m)

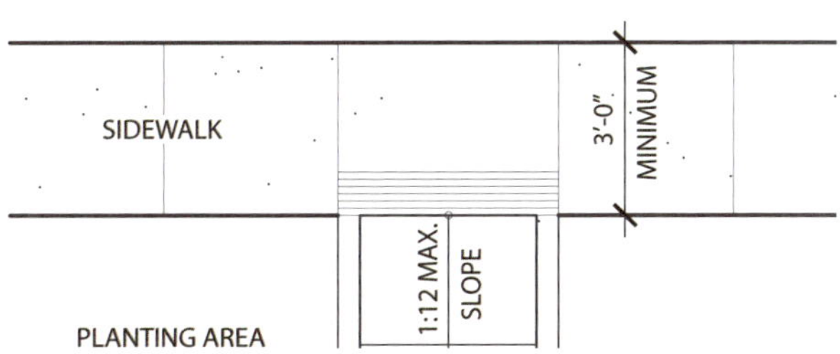

Fig. 2.5—Curb ramp with returned curbs (Note: 1 ft = 0.305 m)

- The slope is required to be measured per Fig. 2.3. Transitions from ramps to walks, gutters, or streets are required to be flush and free of abrupt changes. Maximum slopes of adjoining gutters, road surface immediately adjacent to the curb ramp, or accessible route are not allowed to exceed 1:20;
- The minimum width for curb ramps is 36 in. (915 mm), exclusive of flared sides (Fig. 2.4). The maximum slope of the flare is limited to 1:10 (Fig. 2.4) where it is not protected by handrails or guardrails;
- Curb ramps with returned curbs may be used where pedestrians would not normally walk across the ramp (Fig. 2.5);
- Built-up curb ramps are required to be located so they do not project into vehicular traffic lanes (Fig. 2.6).

Ramps

Any part of an accessible route with a slope greater than 1:20 must meet the ADA requirements for a ramp. These include:

- The least possible slope is required to be used for any ramp. The maximum slope of a ramp in new construc-

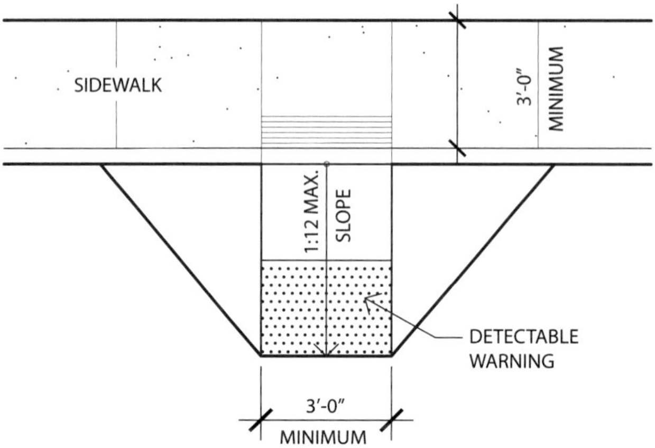

Fig. 2.6—Built-up curb ramp. Note: Not allowed to project into vehicular path (Note: 1 ft = 0.305 m)

tion is 1:12, which is 8.33% or 1 in./ft (83 mm/m). The maximum rise, regardless of length, for any run is 30 in. (760 mm) (Fig. 2.7);

- A minimum clear width of 36 in. (915 mm) (Fig. 2.7);
- Landings are required at the bottom and top of each ramp and each ramp run. Landings are required to have the following features (Fig. 2.7):
  - The landing is required to be at least as wide as the widest ramp run leading to it;
  - The landing length is limited to a minimum of 60 in. (1525 mm) clear; and
  - If ramps change direction at landings, the minimum landing size is 60 × 60 in. (1525 × 1525 mm);
- A cross slope for ramp surfaces no greater than 1:50 is required;
- Ramps and landings with drop-offs are required to have curbs, walls, railings, or projecting surfaces that prevent people from slipping off the ramp; and
- Outdoor ramps and their approaches are required to be designed so that water will not accumulate on walking surfaces.

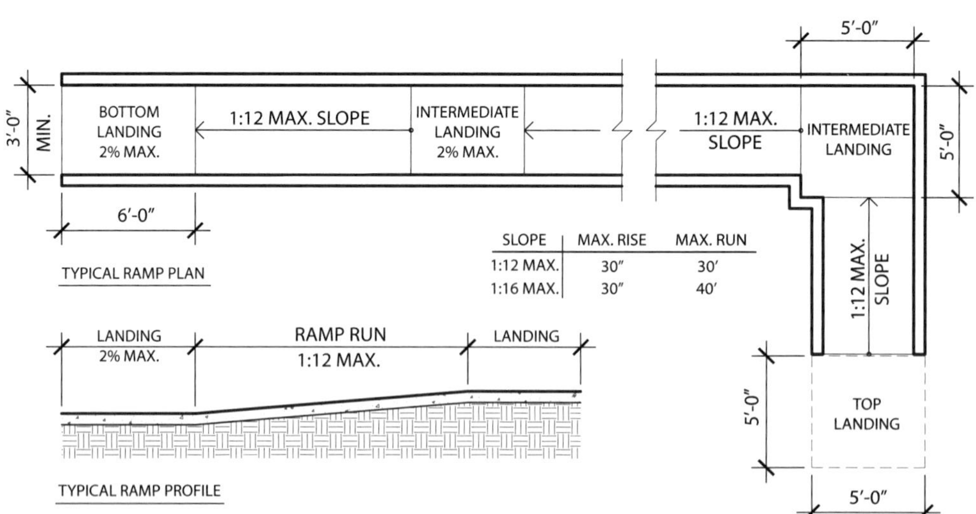

Fig. 2.7— Typical ramp plan and profile (Note: 1 in. = 25.4 mm; 1 ft = 0.305 m)

Stairs

ADA requirements for stairs include:

- On any given flight of stairs, all steps are required to have uniform riser heights and uniform tread widths. Stair treads can be no less than 11 in. (280 mm) wide, measured from riser to riser. Risers are not allowed to exceed 7 in. (177 mm) in height (Fig. 2.8). Open risers are not permitted;
- The undersides of nosings are not allowed to be abrupt. The radius of curvature at the leading edge of the tread can be no greater than ½ in. (13 mm) (Fig. 2.9). Risers are required to be sloped or the underside of the nosing is required to have an angle not less than 60 degrees from the horizontal. Nosings can project no more than 1.5 in. (38 mm) (Fig. 2.8).
- Outdoor stairs and their approaches are required to be designed so that water will not accumulate on walking surfaces.

Detectable Warnings on Walking Surfaces

Detectable warnings (Fig. 2.10) are required to consist of raised, truncated domes with a base diameter of 0.9 to 1.4 in. (23 to 36 mm), a nominal height of 0.2 in. (5 mm), and a nominal center-to-center spacing of 1.6 to 2.4 in. (41 to 61 mm) and are required to contrast visually with adjoining surfaces, either light-on-dark or dark-on-light.

Although detectable warnings are not required in the 2010 ADA guidelines, except on transit platforms, the local building code may contain additional requirements.

Summary

An accessible route is typical at virtually all new construction, including retail centers, office buildings, plazas, parks, public sidewalks, transportation facilities, and residential buildings.

The maximum cross slope on an accessible route is 2% or 1:50 or slightly less than ¼ in./ft (20 mm/m). For

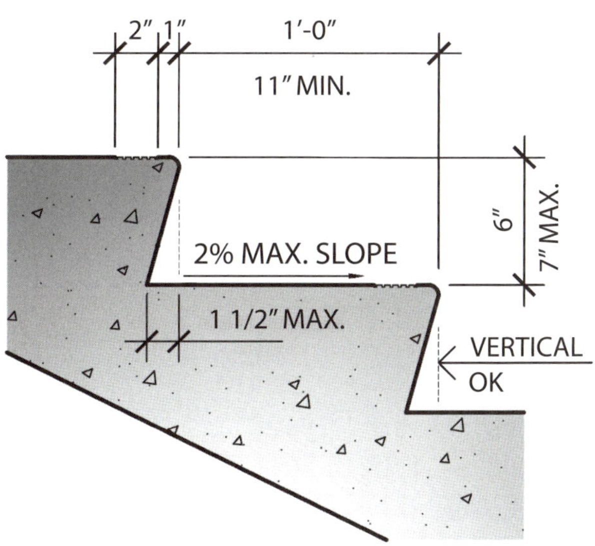

Fig. 2.8—Requirements for stairs include 11 in. (280 mm) minimum treads, 7 in. (177 mm) maximum risers, a nonslip surface, a grooved strip with contrasting color, and ½ in. (13 mm) maximum radius nosing (Note: 1 in. = 25.4 mm)

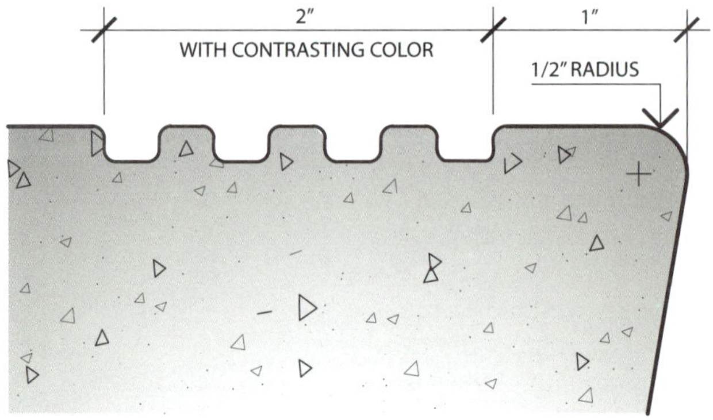

Fig. 2.9—Tooled warning strip (Note: 1 in. = 25.4 mm)

example, you cannot have more than 1 in. (25 mm) cross fall (side to side) on a 50 in. (1.25 m) wide walk.

The maximum running slope (fall in the direction of travel) is 8.33% or 1:12. Therefore, you cannot have more than 1 in. (25 mm) fall in any given 12 in. (300 mm) length of walk.

Any part of an accessible route with a slope greater than 5% or 1:20 is considered a ramp.

Keep in mind that the term "maximum" is exactly that. It is preferred to stay conservatively under the maximum percentages so as not to go even slightly beyond the maximum, otherwise your work may be subject to corrective repairs, removal, or a lawsuit.

Contractors should also consider tolerances when constructing ramps. It is best to construct the ramp below the maximum allowable slope to accommodate tolerances within the specifications. Therefore, a ramp with a maximum slope of 8.33% might best be constructed at 8.00%, or less, to allow for tolerances during placement and finishing of concrete. Contractors should address this issue early through the request for information (RFI) process to gain clarification and steer clear of any violations of ADA Guidelines.

Fig. 2.10—Precast concrete truncated dome tiles at a handicapped access ramp

REFERENCES

1. *2010 ADA Standards for Accessible Design*, Department of Justice, Washington, DC, Sept. 15, 2010, 275 pp.

CHAPTER 3: SOILS, SUBGRADE, AND BASE REQUIREMENTS

Soils Report

If available, review the soils report (also known as a geotechnical investigation report) for the recommendations contained therein. The report has three purposes: (1) evaluate geotechnical conditions at the site; (2) identify significant geotechnical or geologic issues that would impact site development; and (3) provide geotechnical recommendations for design and construction of the proposed development, or more specifically, the concrete work.

Typically, concrete is referred to as a rigid pavement in the report. The soils report will often provide sections and designs for pedestrian and vehicular areas, driveways, fire lanes, structures, walls, footings, curb and gutter, and other concrete flatwork.

The soils report should be reviewed for items that may affect the concrete sitework, including subgrade preparation and cement type and corrosion protection:

Subgrade Preparation

Areas planned for concrete pavement or flatwork (sidewalks) are often overexcavated to a defined depth (for example, 12 or 24 in. [300 or 600 mm]) below the existing or proposed subgrade, whichever is deeper. The excavated soil is reviewed and possibly tested by the geotechnical firm to determine if it is suitable for use as compacted structural fill. If it is determined to be suitable, then the fill soil or import soil, if necessary, is placed in thin, loose lifts, moisture-conditioned as necessary to near optimum moisture content, and recompacted to 90 or 95% relative compaction per the soils report.

If another contractor trenches through the prepared, compacted subgrade (for an electrical conduit, plumbing, or irrigation work, for example), verify that their backfill has been compacted, tested, and accepted by the geotechnical firm's on-site inspector. You do not want the subgrade to fail under your paving at a later date.

A concrete contractor is wise to exclude subgrade preparation from their proposal. Recompaction of existing on-site soil can be difficult to achieve under some circumstances and is best left to a grading contractor.

Cement Type and Corrosion Protection

The geotechnical firm will evaluate the soils through laboratory testing to determine what steps need to be taken to ensure slabs-on-ground are not subject to chemical attack from the underlying soils. The engineer will determine cement type (I/II or V) and the water-cementitious materials ratio (*w*/*cm*) required to protect the concrete.

Other Information

Slabs-on-ground should be placed on properly compacted subgrade as described within the soils report. In addition, the report may recommend the following:

- Concrete thickness;
- Aggregate base thickness;
- Reinforcing steel;
- Compressive strength;
- Water-cementitious materials ratio;
- Supplementary cementitious materials required to protect the concrete from the sulfate-rich or low-pH soils; and
- Recommended joint (contraction joint) spacing, depth of contraction joints (typically one-fourth the depth of the slab) for each type of paving.

For example, pedestrian paving, alleyways, and concrete pavements (heavy truck) may each have their own design. Often, a table is contained within the soils report showing the structural sections required for various loadings.

It is highly recommended that the concrete contractor compare the recommendations in the soils report to the drawings and specifications. It is best to adhere to the most stringent design required. All conflicts in the documents, however, should be disclosed to the Owner, General Contractor, or Contracting Agency.

CHAPTER 4: FORMING, REINFORCEMENT, DOWELS, AND JOINTING

Forming is a critical task for decorative concrete paving. Forming creates the desired layout, sets the elevations for the paving, and determines the placement sequence.

Forming for decorative concrete paving typically consists of using standard sized lumber: 2x4 (38 x 89 mm) lumber for 4 in. (100 mm) thick concrete, 2x6 (38 x 140 mm) lumber for 6 in. (150 mm) thick concrete, and 2x8 (38 x 184 mm) lumber for 8 in. (200 mm) thick concrete. The lumber is held in place by stakes. The stakes are typically made of wood or metal. The choice of material depends on the firmness of the subgrade and the forming system that a contractor chooses.

Typically, the location of the forming is dictated by the layout shown on the plans. The contractor should follow the dimensions on the plan where shown and not scale the drawings. If a dimension cannot be determined from the plans, the contractor should request a clarification from the General Contractor or design professional (Architect, Landscape Architect, or Engineer). In addition to the perimeter forming, there is often interior forming to provide planters, blockouts, banding, or expansion joints.

Placing in lanes provides accessibility and space for multiple operations.

Before forming can begin, the decorative concrete contractor should consider:

- Are adequate dimensions shown on the drawings for layout and forming purposes?
- Are elevations shown on the drawings? If so, do they provide adequate slope to drains and is the paving sloping away from structures? The recommended minimum slope for textured concrete paving is 1% slope. Any slope less than this can create ponding water or "birdbaths".
- Do the slopes on the plans conform to ADA requirements?
- The sequence of placing concrete will determine the forming. Typically, the perimeter forming is placed, followed by the interior forms. The interior forms may be for banding, construction joints, full-depth expansion joints, or where dowels are required. Furthermore, the interior forms may be required due to a change in color or texture, to allow production, or for accessibility purposes.

A contractor needs to proactively determine problems or conflicts within the plans and specifications before forming and certainly before concrete is placed. For example, if the given elevations on a plan cause the crossfall of a sidewalk to exceed 2%, then the contractor must let others know that this is a violation of ADA and the elevations need to be revised. Once the concrete has been placed, it is often too late to raise an issue with the plans or specifications.

There are often dimension or elevation "busts"—that is, inconsistencies that do not work with the on-site conditions, particularly when new construction is tying into existing construction. Again, the contractor needs to identify the problems, preferably before beginning work. Problems with the plans and specifications are typically resolved through the request for information (RFI) process, wherein a contractor states the conflict on a written form and the responsible party answers the RFI before work is begun.

Allowance for Accessibility and Production

In the past, concrete paving was constructed in a checkerboard pattern, instead of in lanes (Fig. 4.1), in an effort to permit the earlier placements to shrink and control cracking. Experience has shown that this is no longer recommended, especially for decorative concrete, because:

- Access is more difficult;
- The joints between the adjacent placements are not as smooth (a saw-cut or tooled contraction joint is preferable to an expansion joint);
- Color and texture uniformity are lessened by the excessive number of placements required; and
- Costs are increased when using a checkerboard pattern.

Fig. 4.1—Forming over a waterproofing membrane. Concrete to be placed in lanes as opposed to a checkerboard

A concrete lane formed between 10 and 20 ft (3 and 6 m) in width is best for decorative concrete. The actual width would depend on the following factors:

- **Type of Finish:** Seeded aggregate will require a closer spacing, whereas a broom finish can be at a wider spacing.
- **Contraction Joints:** The plans may call for saw-cut contraction joints at 10 ft (3 m) on center each way. This would require that the concrete placement lane be 10 or 20 ft (3 or 6 m) wide—that is, a multiple of 10 ft (3 m).
- **Decorative Joints:** The plans may call for the paving to receive a saw-cut or tooled pattern with a given spacing. For example, a 3 x 3 ft (0.9 x 0.9 m) joint pattern would require the paving lane to be divisible by 3, resulting in a lane width of 12, 15, or 18 ft (3.6, 4.5, or 5.4 m).

Fig. 4.2—Light gray concrete banding with dark gray concrete fields at a motor court

Reinforcement

Decorative concrete paving (Fig. 4.3) is typically reinforced with steel. The type, size, and spacing of the reinforcing steel is usually called out on the drawings, specifications, and soils report. Reinforcement for decorative concrete paving consists of welded wire reinforcement, steel reinforcing bars, fibers, or a combination of these.

Welded Wire Reinforcement

Welded wire reinforcement (Fig. 4.3) consists of pieces of wire welded together to form a grid pattern. The size of the grid pattern used in decorative concrete paving is typically 4 x 4 in. up to 8 x 8 in. (102 x 102 mm up to 203 x 203 mm). Because the wires are welded in a mat, the wire spacing does not change when concrete is placed. Common wire sizes are W1.4 (MW9), W2.9 (MW19), and W4.0 (MW26), while common spacing is either 4 or 6 in. (102 or 152 mm). However, welded wire reinforcement is available in many combinations of wire size and spacing.

Most often, 6x6-W1.4xW1.4 or 6x6-W2.9xW2.9 (152 x 152-MW9 x MW9 or 152 x 152-MW19 x MW19) is specified for decorative concrete paving. The first two numbers, 6x6 (152 x 152), denote the spacing of the cross wires and long wires, which in this callout is 6 in. (152 mm) on center each way. The second set of numbers call out the wire size, W1.4 or W2.9 (MW9 or MW19). Depending on the spacing and size of wire, wire reinforcement is available in rolls or mats.

Fig. 4.3—Welded wire reinforcement placed on concrete chairs. Mesh is W4 (MW26) wire on 4 in. (100 mm) centers

Welded wire reinforcement, especially in rolls, presents a challenge to control the location of the mesh during concrete placement. In addition, care must be taken to support the welded wire reinforcement, as necessary, to keep it from being pushed down into the ground. Heavier gauge wire, in mats (not rolls), can more easily be supported during the placement of concrete.

A contractor may be held liable if welded wire reinforcement, or reinforcing steel bar, is not properly positioned in the concrete paving.

Reinforcing Steel Bar

Reinforcing steel bars are usually stocked in 20 ft (6 m) lengths. The bars are typically deformed. That is, a pattern is rolled onto the surface of the bars that allows the concrete to grip the bar. There are several grades of reinforcing steel available, but Grade 60 (420) is by far the most common. Grade 60 has a minimum yield strength of 60,000 psi (420 MPa). The size designations up through No. 8 (No. 25) are the number of eighths of an inch (millimeters) in the nominal diameter of a plain round bar having the same weight per unit length as the deformed bar. For example, the three most common bar sizes for decorative concrete paving are the No. 3, No. 4, and No. 5 (No. 10, No. 13, and No. 16). The No. 3 bar has a 3/8 in. nominal diameter, the No. 4 bar has a 4/8 in. or ½ in. nominal diameter, and the No. 5 bar has a 5/8 in. nominal diameter (the No. 10 bar has a 10 mm nominal diameter, the No. 13 bar has a 13 mm nominal diameter, and the No. 16 bar has a 16 mm nominal diameter).

The size and spacing of the reinforcing steel bars are typically called out on the drawings. However, the specifications and soils report may contain this information as well. Pedestrian paving is typically 4 in. (100 mm) thick concrete, and it can be found to be reinforced with 6x6-W1.4xW1.4 (152 x 152-MW9 x MW9) welded wire reinforcement, No. 3 (No. 10) bars at 24 in. (600 mm) on center each way, or No. 3 (No. 10) bars at 18 in. (450 mm) on center each way. Each project is unique, and a contractor should not assume steel reinforcement based on a previous project. Vehicular paving is typically 6 in. (150 mm) thick, and it is often reinforced by welded wire reinforcement with a larger gauge wire such as 6x6-W2.9xW2.9 (152 x 152-MW19 x MW19), or larger bars at closer spacings. For example, No. 4 (No. 13) bars at 18 in. (450 mm) on center each way or No. 4 (No. 13) bars at 12 in. (300 mm) on center each way.

Spacer Blocks

Welded wire reinforcement and steel reinforcement need to be held in position during concrete placement. Typically in a lightly loaded slab, the concrete reinforcement is contained in the top half of the concrete paving. A 4 in. (100 mm) thick slab should employ the use of 2 x 2 x 2 in. (50 x 50 x 50 mm) concrete spacer blocks or other chairing devices, whereas a 6 in. (150 mm) slab uses 3 x 3 x 3 in. (75 x 75 x 75 mm) concrete spacer blocks (Fig. 4.4). The blocks are spaced as needed to hold the reinforcement at the proper height and provide adequate support during placement of the concrete.

Fig. 4.4—Reinforcing steel bar placed on 3 in. (75 mm) thick precast concrete spacer blocks for 6 in. (150 mm) thick paving

Dowels

Dowels (Fig. 4.5) are used between adjacent concrete placements to allow the slab to move horizontally in the same longitudinal direction as the dowel without restraint. Dowels are a means to allow for load transfer across construction or control joints. Dowels are typically placed at 12, 18, or 24 in. (300, 450, or 600 mm) on center, depending on the loading requirements. In the past, smooth steel dowels were placed in the slab through holes drilled in the edge form. After the form was stripped, the end of the smooth dowel was either greased or placed in a cylindrical cardboard sleeve. In this way, the dowel was fixed on one end and allowed to move on the other end, hence the term "slip" dowel. This allowed horizontal movement but restricted the vertical movement of the slabs.

Today there are time-saving products available for dowels. One system uses plate dowels, either square or diamond shaped, encased in plastic sleeves. The plastic sleeve is attached to the edge form with nails prior to concrete being placed. Once the form is stripped, the plate dowel is inserted into the void created by the plastic sleeve. The use of these products in exterior applications should be evaluated for possible issues in environments subject to cycles of freezing and thawing. Another system uses a round or square dowel, plastic base, and sleeve attached to the edge form prior to concrete placement. The form and base are removed after placing and the steel dowel is inserted into the sleeve that is encased in concrete. Both product types claim to reduce installation costs by 50%. They both eliminate drilling of the forms and allow the slab to move horizontally while restricting the movement vertically.

Fig. 4.5—Expansion joint material, steel dowels, and plastic sheeting for protection; ready for a concrete placement

Dowels are essential between adjacent concrete placements to restrict vertical movement. Concrete paving that moves vertically can create a tripping hazard, cause water to pond, and may be cause for removal and replacement. It is very important that dowels are placed perpendicular to the joint and level so as to not cause restriction in horizontal slab movement.

Microfibers and Macrofibers

Microfibers and macrofibers are secondary reinforcement. If used, microfibers are more common in decorative concrete because they are less visible in the concrete than macrofibers. Neither are meant to replace welded wire reinforcement or reinforcing bars. Microfibers may reduce cracking due to plastic shrinkage and thermal expansion and contraction, reduce permeability and absorption, and increase durability of concrete. Fibers are added directly to the concrete mixture either at the batch plant or at the jobsite. When added at a jobsite, the concrete needs to be mixed at a high speed during addition of the fibers and then at mixing speed for a minimum of 40 revolutions, after the addition of fibers. Consult with the fiber manufacturer to verify specific requirements. The dosage rate for fibers typically falls in a range of 0.5 to 2.0 lb/yd^3 (0.3 to 1.2 kg/m^3) depending on the application and the manufacturer. Fibers do not require additional water or admixtures when they are added to the concrete.

Decorative concrete typically uses finer, shorter fibers, ranging in length from ¼ to ¾ in. (6 to 19 mm) depending on the manufacturer, to minimize the appearance of the fiber in the finished concrete product, whether that is paving, curbs, stairs, or walls. The fibers will typically be visible in the final product. Therefore, they may need to be burned off with a torch.

Jointing

Isolation Joints

Isolation joints (also known as expansion joints) and contraction joints (also known as control joints) are placed in decorative concrete paving to minimize random shrinkage cracking. Isolation joints should be minimized. They are typically placed against structures, walls, existing concrete, and other vertical restraints. They can be placed at re-entrant corners to reduce cracking with more certainty than a contraction joint. All re-entrant corners should have an isolation, construction, or contraction joint. Where this is not possible, a minimum of two No. 4 reinforcing bars 4 ft (1.2 m) in length should be placed on a 45-degree angle to the corner. Isolation joints should not be used to create a checkerboard placement–this is not necessary.

Contraction Joints

Contraction joints are placed in concrete to minimize shrinkage cracking. The joints are spaced ideally to form square panels. Concrete that is 4 in. (100 mm) thick should have contraction joint spacing no more than 10 ft (3 m) on center, while 6 in. (150 mm) thick concrete should have contraction joints no more than 15 ft (4.6 m) on center. The longer dimension of a concrete panel should not exceed the shorter dimension by more than 50%. For example, a 4 in. (100 mm) thick panel could be 7 x 10 ft (2.1 x 3 m). However, if a panel is 5 ft (1.5 m) wide, the joints should be less than 10 ft (3 m) apart or they will probably crack on their own. Always try to be as conservative as possible. A concrete mixture with a lower water-cementitious materials ratio will generally have less shrinkage cracking. In addition, a properly prepared, compacted, and leveled subgrade will aid in minimizing shrinkage cracking as it provides a low coefficient of friction as the slab contracts during curing.

Cracking will be uncontrolled unless contraction joints are placed in the slab similar to Fig. 4.6 and 4.7.

Residential driveway and front entry

A joint layout for a residential driveway and front entry is illustrated in Fig. 4.6. The concrete fields are placed in two separate placements. In this example, they are saw-cut

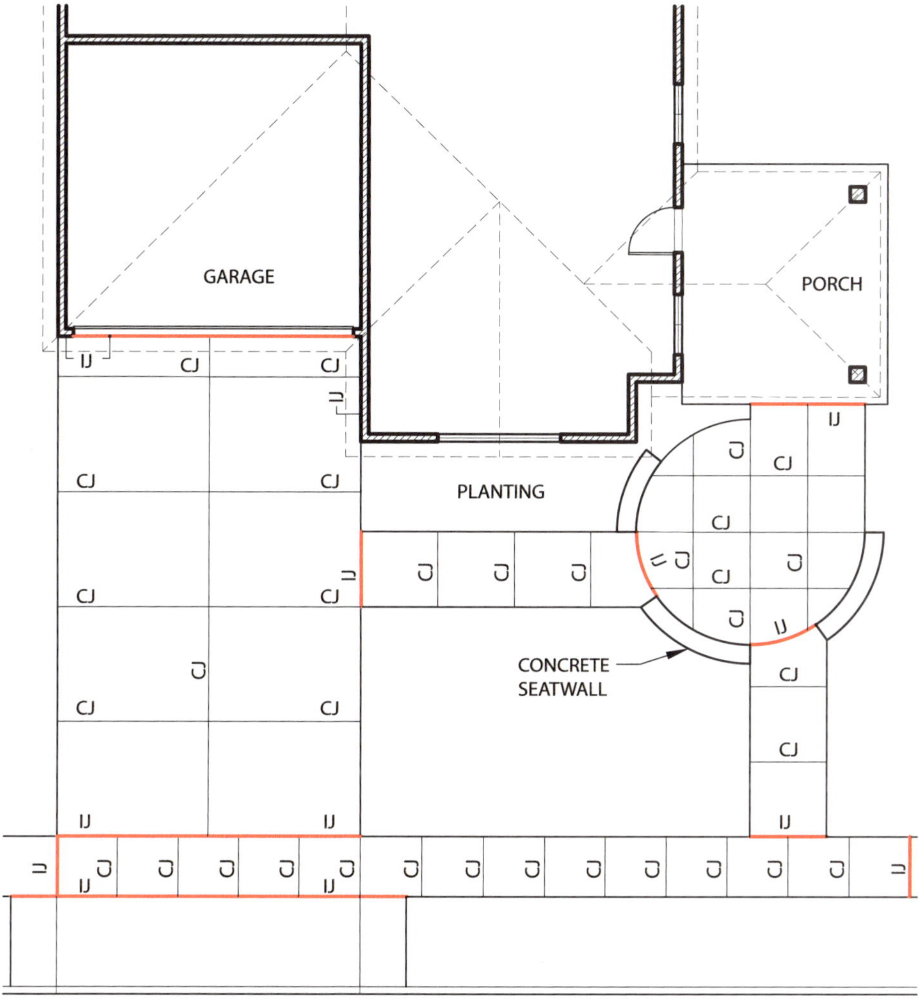

Fig. 4.6—Residential driveway and front entry jointing plan

(or tooled) to create equal panels and minimize shrinkage cracking. A ¼ in. (6 mm) thick isolation joint material is placed against the residence, seatwalls, driveway approach, and other vertical restraints prior to concrete placement. The top portion of this material will be removed after the concrete paving fields are placed and the void will be caulked. Steel dowels are placed between concrete placements at 24 in. (600 mm) on center to minimize vertical movement between the paving and banding.

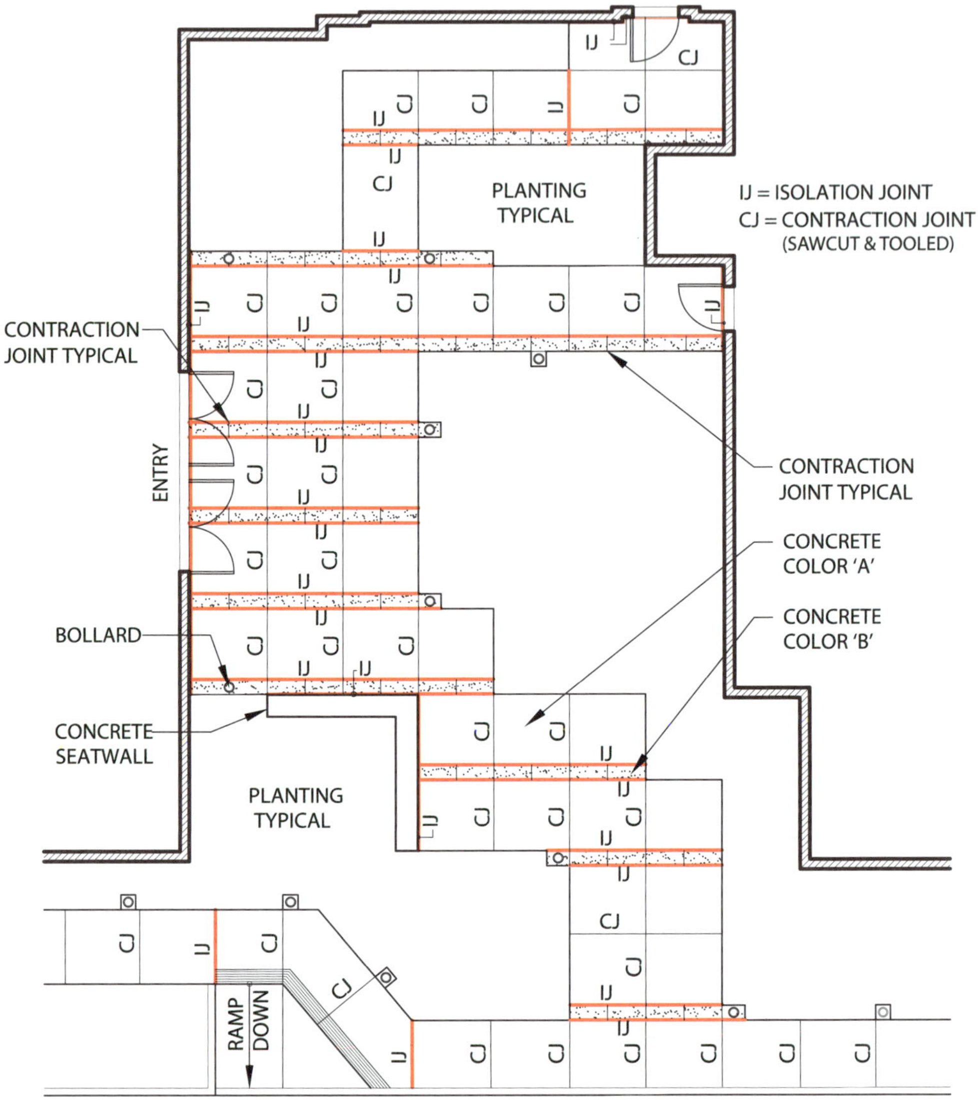

Fig. 4.7—Commercial project jointing plan

Commercial Project

An example of a joint layout for a commercial project is shown in Fig. 4.7. The concrete banding is placed first. The 12 in. (300 mm) wide bands are saw-cut (or tooled), to create contraction joints, at approximately 36 in. (0.9 m) on center. A ¼ in. (6 mm) thick isolation joint material is placed against the banding. The top portion of this material will be removed after the concrete paving fields are placed and the void will be caulked. Steel dowels are placed between the banding and the paving at 24 in. (600 mm) on center to minimize vertical movement between the paving and banding.

Isolation joint material is placed against the building, seatwalls, curbs, and other vertical restraints prior to concrete placement. In this example, the concrete fields are saw-cut (or tooled) 6 ft (1.8 m) on center, or to create equal panels. This will help minimize shrinkage cracking. Contraction joints are always better than cracks.

Joints between Adjacent Concrete Placements

Isolation joints permit horizontal and vertical movement between the concrete paving and any walls, columns, vertical restraints, or adjacent concrete placements. Concrete shrinks as it cures. If the slab is rigidly joined to a fixed object, the slab is likely to crack because the drying shrinkage cannot be accommodated.

Isolation joint material, the full depth of the concrete paving edge, is placed between concrete placements and against walls, columns, and vertical restraints. Isolation joints can be composed of various materials, but decorative concrete contractors rarely use asphalt-impregnated fiber. Foam, with a perforated top for caulking purposes, is typically placed between concrete placements. The foam ranges from ¼ to ½ in. (6 to 12.5 mm) thick. Refer to the following sections for a discussion of joint details.

Saw-cut Isolation Joint

In the sawcut isolation joint detail shown in Fig. 4.8, the concrete slab in Placement 1 is placed and edged using no more than a ⅛ in. (3 mm) radius edger. Next, ¼ in. (6 mm) thick foam material is glued in place using spray glue. It is left 1 in. (25 mm) below the top of the concrete. It must be straight and flush with the face of Place-

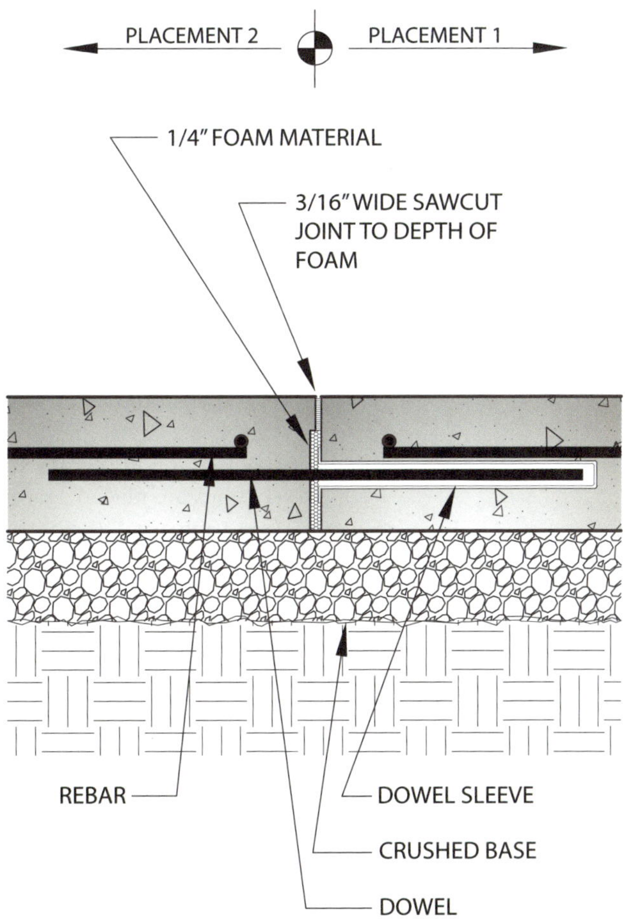

Fig. 4.8—Saw-cut isolation joint detail (Note: 1 in. = 25.4 mm)

ment 1. Concrete in Placement 2 is placed and finished over the top of the expansion joint material with a ⅛ in. (3 mm) radius edger. The saw-cut joint, extending down into the foam, is sawed when the concrete has hardened sufficiently. The concrete in Placement 2 will be cantilevering over the foam. It is this section of concrete that must be cut out using a 3/16 in. (5 mm) wide blade.

Isolation Joint with Joint Sealant

In the detail for an isolation joint with joint sealant shown in Fig. 4.9, the concrete slab in Placement 1 is placed and edged with a concrete edger per plans. If no edger is specified, a ¼ in. (6 mm) radius is recommended. Next, ¼ or ½ in. (6 or 12.5 mm) thick foam material is glued in place using spray glue or nailed in placed using concrete nails. The isolation joint material must be straight and flush with the face of Placement 1. Concrete in Placement 2 is placed against the foam expansion joint material. Care must be taken not to get any concrete behind the foam. If this does occur, it should be removed immediately using a margin trowel.

Saw-cut Construction Joint

In the detail for a saw-cut construction joint shown in Fig. 4.10, a saw cut is made where concrete placement has concluded for the day, usually conforming to a predetermined joint layout. This joint should receive dowels. Dowels permit horizontal movement, but little vertical movement. Doweled joints (Fig. 4.11) have generally replaced keyed joints, as keyed joints shrink and often fail under heavy loading conditions. In this detail, the two concrete paving sections are placed against one another and edged with a 1/8 in. (3 mm) radius edger. Next, a saw cut is placed directly over the construction joint to remove a section of the concrete approximately 3/16 x 1 in. deep (5 x 25 mm deep). This makes a more aesthetically pleasing construction joint that will tie into the adjacent saw-cut concrete paving.

Contraction Joints

To reduce the occurrence of random cracks, contraction joints are used to create straight-line planes of weakness in the slab. As the slab shrinks and tends to curl, the joints open slightly and cracks occur at the predetermined locations instead of randomly over the slab. Depth of tooled or saw-cut joints should be ¼ the slab thickness, *T*. The planes of weakness may be established by saw cuts after the concrete is finished and has hardened, or by using jointing tools to yield a joint similar to the tooled joint shown in Fig. 4.12.

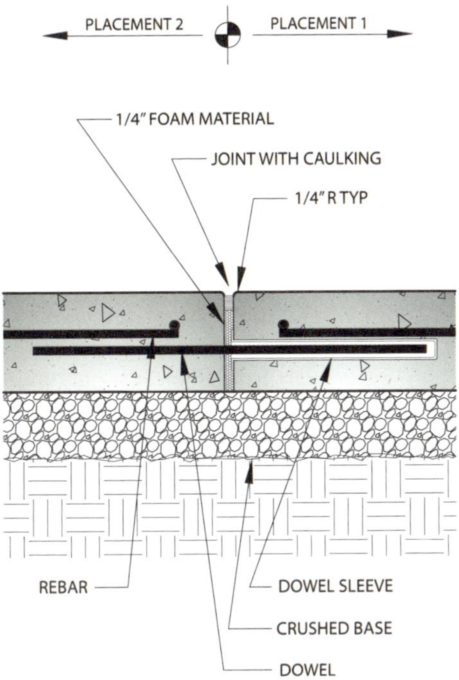

Fig. 4.9—Isolation joint with joint sealant detail (Note: 1 in. = 25.4 mm)

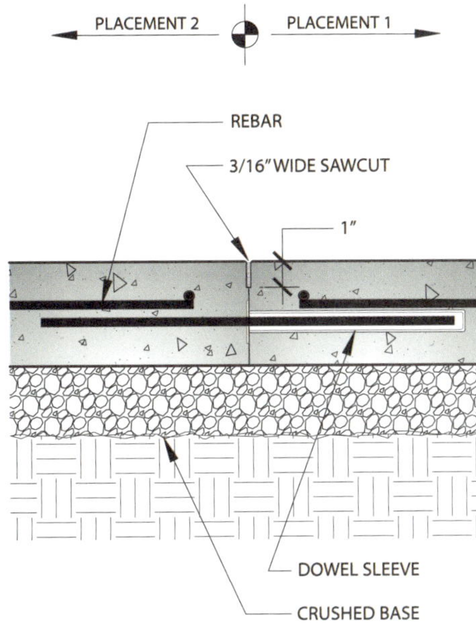

Fig. 4.10—Saw-cut construction joint detail (Note: 1 in. = 25.4 mm)

(a)

(b)

Fig. 4.11—Sleeves can be used to position dowels in concrete slabs: (a) a dowel sleeve in existing concrete; and (b) a steel dowel inserted through the isolation joint material into the dowel sleeve

Tooled Joints

Tooled joints require jointing tools manufactured with the specified radius and depth of groove. ACI 302.1R-04,[1] "Guide for Concrete Floor and Slab Construction," recommends a 1/8 in. (3 mm) radius.

Saw-cut Joints

Saw-cut joints (Fig. 4.13) should be sawed as soon as the concrete is hard enough not to be torn or damaged by the blade, but before random cracks can form in the concrete slab. This requires sawing the slab from 4 to 24 hours after finishing is complete. The timing depends on air temperature, concrete mixture, thickness of the concrete slab, type of saw, and the substrate. Saw-cut joints at the earliest possible opportunity when a quality result can be provided.

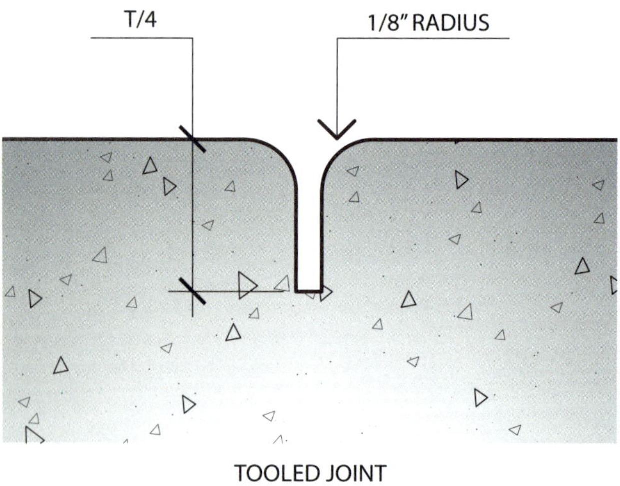

Fig. 4.12—A tooled joint (Note: 1 in. = 25.4 mm)

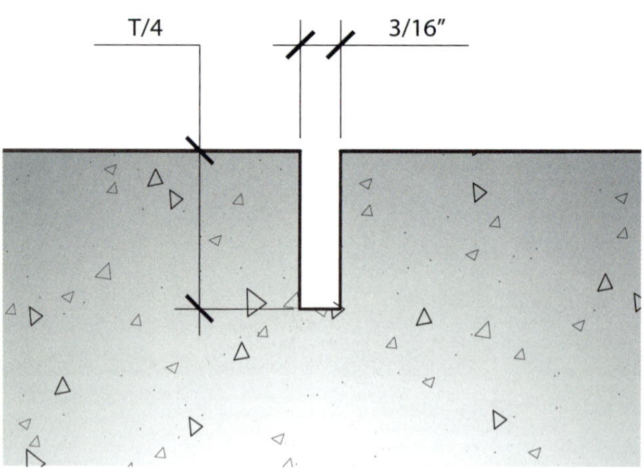

Fig. 4.13—A 3/16 in. (5 mm) wide saw-cut joint (Note: 1 in. = 25.4 mm)

REFERENCES

1. ACI Committee 302, "Guide for Concrete Floor and Slab Construction (ACI 302.1R-04)," American Concrete Institute, Farmington Hills, MI, 2004, 76 pp.

CHAPTER 5: BASIC DECORATIVE CONCRETE COMPONENTS

Mixtures for Decorative Concrete

When ordering concrete for use in placing decorative concrete flatwork, you can specify several properties of both the materials used in the concrete and the concrete itself. These include:

- Nominal maximum size of the aggregate;
- Type of portland cement and other cementitious materials;
- Amount of cementitious material per cubic yard (cubic meter), or compressive strength;
- Air content for air-entrained concrete;
- Concrete slump;
- Type and amount of pigment for integrally colored concrete;
- Admixtures, such as shrinkage-reducing and water-reducing admixtures; and
- Type and amount of fibers.

Another approach to ordering concrete is to specify the performance properties of the concrete rather than the materials and proportions to be used. The two approaches, however, should not be mixed.

Nominal Maximum Size of Aggregate

The larger the nominal maximum aggregate size, the lower the water requirement to produce a given slump. Because higher water contents increase concrete shrinkage—and the possibility of random cracks—it's best to use as little water as possible in the concrete while still getting the needed slump. That means using the largest possible nominal maximum size aggregate. The most commonly used nominal maximum aggregate size for decorative concrete is ¾ in. (19 mm) aggregate. When tooled joints are required or stamped, concrete patterns exceed ¼ in. (6 mm) in depth; 3/4 in. (19 mm) aggregate may be too large and concrete made with ½ in. (13 mm) nominal maximum size aggregate will be needed. Finally, when the concrete will be pumped through small-line pumps with a hose diameter of 2 or 3 in. (50 or 75 mm), 3/8 in. (9.5 mm) nominal maximum size aggregate or pea gravel may have to be used. Pea gravel mixtures that have high water-cementitious materials ratios are likely to shrink and crack more than concretes made with larger coarse aggregate; as a result, they can be problematic for decorative concrete work. Pumping concrete with larger aggregate sizes requires use of a more expensive 4 in.

(100 mm) diameter line, but the concrete quality will generally be better.

Concrete mixtures that contain 70% sand and 30% pea gravel are an example of concrete mixtures that will exhibit a large amount of shrinkage due to a relatively high water-cementitious materials ratio. Contractors often use such mixtures for ease of placement, but the durability and structural capacity of the concrete is compromised.

Portland Cement and Other Cementitious Materials

Portland cements are the most common cementitious materials used in decorative concrete. They are hydraulic cements, which means they set and harden by reacting with water, and they are not soluble in water after hardening. They come in several types, but the most commonly used in decorative concrete are Type I and Type II. Type I is a general-purpose portland cement suitable for all uses where the special properties of other types are not required. Type II portland cement has a moderate resistance to sulfate attack and is used when soils and ground water in contact with the concrete have higher concentrations of sulfate than normal but are not unusually severe. Type II cement is made with less of the compound that is attacked by sulfates, and it also puts off less heat when it reacts with water. Because of its increased availability, Type II cement is sometimes used for all concrete construction in an area, regardless of the need for sulfate resistance or moderate heat generation. Some cements may be labeled with more than one type designation, for example Type I/II or Type II/V. This simply means that such a cement meets the requirements of both cement Types I and II or Types II and V. Most cements are a shade of gray.

In areas with very high sulfate concentrations, Type V cement—which has a higher sulfate resistance than Type II cement—may be needed and should be used in concrete that also has a lower water-cementitious materials ratio, as shown in ACI 318-14,[1] Table 19.3.2.1. This will typically result in a higher-strength concrete than is normally used for decorative concrete work.

Some other cement options include using white cement or Type III cement. White cement combined with light-colored aggregates produces more intense colors for either integrally colored concrete or concrete colored with dry shakes. Type III cement produces higher early strengths and is especially useful in cold-weather concrete placing. Type III or white cements cost more than Type I cement, but are sometimes used to get early strength or the color desired by the customer.

Cement truck

Other cementitious materials, referred to as supplementary cementitious materials (SCMs), may be used in decorative concrete. Each should be evaluated for its effects on the finished appearance and performance during and after placement. Common SCMs include fly ash and slag cement. Fly ash is the very fine by-product of burning coal in power plants. It is a pozzolan, which means it has cementitious properties when combined with calcium hydroxide and water. Calcium hydroxide forms in the pore solution of the concrete matrix as part of the reaction between portland cement and water, so when fly ash is added as a concrete ingredient, it can increase the concrete strength by combining with the calcium hydroxide. Adding fly ash to a given concrete mixture may increase the workability, pumpability, and durability of the concrete.

Slag cement is made by grinding granulated blast furnace slag to a very fine powder. Like portland cement, slag cement is hydraulic, so when combined with water, it reacts and hardens. Properly designed and placed mixtures containing fly ash or slag cement in place of some of the portland cement can have increased durability. The manufacture of portland cement generates greenhouse gas emissions, so replacing portland cement with an SCM will reduce the greenhouse gas emissions associated with a given mixture. The strong environmental movement has resulted in strong pressure on producers to use more fly ash and slag cement. However, this can have some harmful effects on decorative concrete due to color variation.

Concrete mixer truck

Fly ashes produced at different power plants or at one plant with different coal sources may have different colors. Fly ash color and the amount used can influence the color of the resulting hardened concrete—making it darker or lighter—in the same way as changes in cement or fine aggregate can change the color. Variability of fly ash color should be monitored by concrete producers who supply concrete for decorative work. Concrete producers can test the fly ash by comparing each delivery to a reference sample. Contractors must rely upon the producer to use fly ash with a consistent color to decrease the variability of their finished work.

Material hauler getting loaded with aggregate

The use of slag cements in a concrete mixture also may darken or lighten the color or produce erratic color variation. Fly ash and slag cement can also slow the setting time of the concrete.

Amount of Cementitious Materials

For many years, concrete cement content was described by referring to the number of bags or sacks of cement used in the concrete. Traditionally, a bag or sack of cement weighed 94 lb (43 kg), so a five-bag mixture contains 470 lb of cement per cubic yard (278 kg of cement per cubic meter) and a six-bag mixture contains 564 lb of cement per cubic yard (335 kilograms of cement per cubic meter). Many contractors still order designated bag mixtures from their concrete producer, but what they usually want is either a 3000 or 4000 psi (21 or 28 MPa) concrete that may contain portland cement, fly ash, and perhaps slag cement. Unless a specific cement content is important to produce a specific appearance or paste content, it may be advantageous for the contractor to order a mixture by specifying strength rather than cement content.

For areas that experience cycles of freezing and thawing, a minimum strength of 4500 psi (31 MPa) and a water-cementitious materials ratio of 0.45 is usually required to produce durable concrete. In warmer climates, 3000 psi (21 MPa) concrete may be adequate as long as there is enough paste (also called "cream") in the concrete to permit good finishing results. Sulfates may require a lower water-cementitious materials ratio, thus resulting in a higher concrete strength. In addition, a structural engineer may specify a higher compressive strength due to loading requirements.

Air Content for Air-Entrained Concrete

Entrained air is generally needed only in concrete that will be exposed to cycles of freezing and thawing and to deicing chemicals. Without entrained air, the concrete is likely to scale or disintegrate. If the concrete will not be exposed to freezing-and-thawing cycles, it is best to use non-air-entrained concrete. The amount of air needed is given as a percent of air by volume of the concrete. If the volume of all the air in a cubic yard of concrete (27 cubic feet) is 1.35 cubic feet, the air content is (1.35/27) × 100, or 5%. The air content needed to protect the concrete increases as the maximum size of the aggregate decreases. Recommended air contents for the three most common coarse aggregate sizes used in decorative concrete are shown in Table 5.1.

Although entrained air improves durability, it can cause problems finishing concrete.

Concrete Slump and Use of Admixtures

Slump is an indicator of the consistency of concrete. Because it was believed for many years that high slumps indicated high water contents, low strengths, and increased shrinkage, the slumps required for concrete flatwork were often set very low—in the range of 2 to 4 in. (50 to 100 mm). Admixtures that increase slump without increasing water content (called water-reducing admixtures) are now commonly used and allow good quality concrete flatwork to be placed at 5 to 8 in. (125 to 200 mm) slumps. That is also the slump range for many decorative concrete flatwork jobs.

Admixtures can also be used to slow the setting of concrete if there is a delay in delivery or a breakdown, or during hot weather when concrete sets faster. These admixtures are called retarding admixtures. Most water-reducing

Table 5.1 – ACI 318-14[1] air content requirements for concrete exposed to freezing and thawing

Nominal maximum aggregate size, in. (mm)*	Air content, percent	
	Concrete exposed to freezing-and-thawing cycles and occasional exposure to moisture	Concrete exposed to freezing-and-thawing cycles and in continuous contact with moisture with or without deicing chemicals
3/8 (9.5)	6	7.5
½ (12.5)	5.5	7
¾ (19.0)	5	6

*Refer to ASTM C33/C33M for tolerance on oversize for various nominal maximum size designations.

admixtures also retard set by up to an hour. Other admixtures speed the setting of concrete, which is helpful in cold weather. These are called accelerating admixtures. One of the most common accelerating admixtures, calcium chloride, affects color and should not be used. Non-chloride accelerating admixtures are available and are recommended for use in decorative concrete.

Shrinkage-reducing admixtures are liquid admixtures that reduce concrete shrinkage caused by drying. They should be used with caution in air-entrained concrete. Shrinkage-reducing admixtures act chemically to reduce some of the internal forces that cause early shrinkage and curling.

When selecting admixtures for use in decorative concrete, attention needs to be paid to the added product costs.

Yellow, black, medium red, and light red integral color liquid pigments

Type and Amount of Pigment for Integrally Colored Concrete

As described later in this chapter, some decorative concrete is colored by adding pigments in powder or liquid form at the batch plant. The pigment dosage is based on the desired intensity of the color. The pigment dosage should not exceed 10% by weight of the cementitious material in the concrete. That's because increasing the dosage beyond this value doesn't improve the color much, but does increase water demand and may have undesirable effects on other concrete properties such as air content, slump, and strength.

Types of Color

Integral Color

Integral coloring materials are a blend of natural or synthetic iron oxide pigments that should conform to ASTM C979/C979M, "Standard Specification for Pigments for Integrally Colored Concrete."[2] These granular, powdered, or liquid products are added to the concrete during mixing operations and prior to concrete placement to produce the desired color. Synthetic oxides, as opposed to natural iron oxide pigments, are usually more uniform in color for longer periods of time.

Carbon black pigments are not color-fast and may reduce the effect of air-entraining admixtures. Unlike other integral color pigments, carbon black is highest in tint strength and the most economical, but it will fade if it is not sealed against water penetration. The use of carbon black in exterior concrete paving should be avoided. Carbon black does not conform to ASTM C979. Always check with the concrete batch plant when ordering black or gray colored concrete for exterior paving use to make sure that the color ordered does not contain carbon black.

Color additives may react chemically with other products used on the surface, such as surface retarders or muriatic acid, and should be tested before use. Various iron oxides produce shades of yellow, buff, tan, brown, red, and gray. The color shade depends on the dosage rate of the pigment used, but most applications produce soft earth tones when combined in a given gray cement-based mixture.

Integral color pigments are sometimes packaged in bags that disintegrate when added to the concrete along with aggregate, cement, and water. These can be added either at the concrete plant, which is usually the preferred method, or at the jobsite. Truck mixers should operate at charging or mixing speeds of 50 to 100 revolutions per minute for 5 to 10 minutes after adding the bags.

Liquid products, with the pigments suspended in water, are also available. Blending four primary shades of synthetic iron oxide permits liquid color manufacturers to produce many different colors.

Integral color costs can vary widely, depending on the desired color saturation (smaller pigment dosages produce lighter colors) and the selected color. For example, a green pigment is considerably higher in cost than a red pigment. Their versatility, however, is superior to that of dry-shakes (to be discussed next). Integral color extends throughout the full depth of the slab, thus allowing any surface texturing method, including a heavy washed or sandblasted surface. Color hardeners are limited to a broomed, stamped, or acid-washed finishes, with sandblasting being risky.

The quantity of pigment is expressed as a percentage of the cementitious materials content by weight. Amounts of pigment in excess of 5% seldom produce greater color intensity, whereas amounts greater than 10% may be harmful to concrete quality.

Close-up of light red and yellow integral color liquid pigments

Integral color pigments formulated for eight unique colors

Pigments will produce more intense colors when used with white or light-colored cements.

Chips or gouges don't expose plain gray concrete when integral color is used, but if patching is needed for large areas or deep defects, color matching is more difficult than matching the color of a dry-shake material.

Dry-Shake Color Hardeners

Good-quality dry-shake color hardeners are mixtures of pigments, gray or white portland cement, graded fine silica sand, and wetting agents. They are packaged in sacks or buckets and are hand broadcast onto the surface of the fresh concrete, then floated into the top surface. Color hardeners produce relatively thin—perhaps 1/8 to 3/16 in. (3 to 5 mm) thick—colored surfaces that are more vibrant than those produced with integral colors. In addition, color hardeners are available in a much wider palette of colors than integral color. Pigments used in hardeners should also meet the requirements of ASTM C979.

Because they contain cement and fine aggregates, color hardeners densify the surface, making it stronger, less permeable, and more abrasion-resistant. The added paste content at the surface also helps produce sharper imprints. In slabs thicker than 4 in. (100 mm), dry-shake color hardeners may be a more economical choice than integral color.

Release Agents

Release agents are hydrophobic (water-repelling) powders used on stamps, stamping mats, or skins to prevent them from sticking to the fresh concrete. They also add an accent color to either integrally colored or dry-shake colored concrete. The accent color is most notable when the base color is light and the release agent is much darker. Most of the release agent does not get embedded in the concrete by the stamping mats or skins. Release agent is meant to remain only in grout lines and lower areas of the stamp pattern; 75 to 80% of all release agent should be removed during the cleaning process to prevent sealer delamination. The remaining embedded powdered release agent creates accent colors that enhance the appearance of many patterns.

Powdered release agents have a downside. They are very fine and become airborne when applied to the surface with a brush or a gloved hand. Workers need to wear masks to prevent them from inhaling the dust, and if the dust is blown around on windy days it can stain buildings, other flatwork, or vegetation. Because it is water-repellent, it remains on the surface of the water and must be contained so as not to let it enter a storm drain system or other body of water.

Although clear liquid release allows the stamp contractor to actively see the progress of their work, it lacks the secondary color we have become used to with powdered release agents. Specialty powdered antiquing agents mixed with water can be used following the surface cleaning, giving the contractor the ability to create a powdered release look without the mess. Antiquing agents can be sprayed on or splashed on the surface with a broom to ensure that the colored liquid remains in the low spots of the surface preventing future sealer problems. The antiquing agent contains a polymer that creates a permanent bond with the surface that exceeds the durability of using powdered release agent mixed with water.

Surface Retarders and Set-Retarding Admixtures

The oldest application of surface retarders was for producing exposed-aggregate finishes with select single-size seeding aggregates. The top of a base slab was struck off to a depth below the top of the forms so adding the seeding aggregate would bring the finished surface to form level. After bull floating the base slab, the seeding aggregate was broadcast by shovel or hand onto the slab surface until the surface was completely covered with a single layer of closely spaced particles. Particles were tamped into the fresh concrete with a wood float, darby, or straightedge and the whole surface was bull floated for final embedment with mortar slightly covering all aggregate particles to approximately a 1/16 in. (2 mm) depth. A surface retarder may have then been sprayed on at this stage to delay exposing the aggregate for several hours. The surface was kept damp by covering it with plastic sheeting. On the same day as the placement, finishers began exposing the aggregate, first by brushing the entire slab surface lightly with a broom to remove excess mortar, then by brushing the surface again and flushing it with a fine spray of water from a hose. Finishers continued brushing and spraying until the aggregate was exposed to the proper depth and the flush water ran clear.

Today, many finishers use set retarders that are able to retard setting of the top mortar layer much longer than a few hours so they can wait until the next day to expose the aggregate. This allows the base concrete longer to harden so the seeded aggregate is less likely to be dislodged during the brushing and flushing operations. Today, surface retarders are manufactured to achieve light, medium, or deep exposures, with several different exposure levels available in each category. Some products are colored, with different colors representing different exposure depths that range from the top of the sand particles to approximately the top 60% of 1½ in. (38 mm) coarse aggregate particles. Top-surface retarders that create a fine sand texture are increasingly being used in states with tough environmental protection laws because these products eliminate acid washing or sandblasting, neither of which is considered to be environmentally friendly.

Because higher temperatures speed setting time, set-retarding admixtures can be used during hot weather concrete placing. Set-retarding admixtures batched at the concrete plant counteract some of the effects of high temperatures and allow more time for finishing the concrete. Prepackaged admixtures can also be added to the concrete truck at the jobsite to slow setting when equipment breakdowns or other unexpected events cause placement delays.

Basic Tools for Decorative Flatwork

Many tools used for decorative concrete are also used for all concrete flatwork. They are used to strike off the concrete; to float, edge, and groove it; and to trowel it. Other tools, such as brooms, produce texture. This is a brief review of the common tools and their uses, with special emphasis where needed on decorative applications. Most of the tools are listed in the order in which they should be used.

Strikeoffs or Straightedges

This tool removes excess concrete and brings the top surface of the concrete to the elevation of the forms or to pads of concrete called wet screeds. Screeding (also called strikeoff) is the first step in finishing concrete and an important step in producing a plane surface. It is done as soon as possible after the concrete is spread, and must be completed before bleedwater appears on the surface.

A piece of 2x4 lumber cut to the correct length—approximately 2 ft (0.6 m) longer than the distance between edge forms or wet screeds—can be used as a straightedge, but many pieces of dimension lumber are not straight and may leave a crown in the surface. For decorative work, it's preferable to use hollow aluminum or magnesium straightedges available from finishing tool manufacturers in 1 x 4, 2 x 4, or 2 x 5 in. (25 x 100, 50 x 100, or 50 x 125 mm) cross sections and 6 to 24 ft (1.8 to 7.3 m) lengths. Wood straightedges are heavier, harder to clean, and may warp after being used.

Straightedges are pulled back and forth with a sawing motion and this can be hard on the worker's hands. One manufacturer produces an extruded magnesium straightedge with a handhold lip for better control and some finishers modify wooden members with handholds. A vibrating screed is the best solution whenever job conditions are such that this tool can be used on the slab. It will eliminate much of the hard work involved in initial leveling of the surface.

Screeding - The first step in finishing concrete

Bull Floats and Darbies

These tools level ridges and fill voids left by striking off the concrete with a straightedge. They also increase the amount of mortar on the surface by slightly embedding the coarse aggregate. Blades of bull floats and darbies are flat and rectangular and can be made of wood, aluminum, magnesium, or composite materials. Designed to float large slab areas, bull floats are usually 8 in. (200 mm) wide and 42 to 60 in. (1 to 1.50 m) long, with a handle 4 to 16 ft (1 to 5 m) in length. Wood bull floats are heavier than those made of other materials and will more easily cut high spots, fill in low spots, and embed the coarse aggregate. But they may tear the surface of air-entrained concrete, which is better finished with bull floats made from other materials.

A bullfloat is used to smooth the concrete after screeding

Darbies are smaller handheld tools designed to float small or restricted slab areas. They measure 3 to 4 in. (75 to 100 mm) in width and 3 to 8 ft (1 to 2.5 m) in length.

Edgers

Edgers densify the concrete surface along a formed perimeter to produce neat, rounded slab edges that are less likely to chip or spall after form removal. Hand edgers range from 3 to 6 in. (75 to 100 mm) wide and 6 to 10 in. (150 to 255 mm) long, with radii from 1/8 to 1 in. (3 to 25 mm). They are made of bronze, or stainless or blue steel. Bronze edgers are heavier, last longer, and produce a smoother, denser edge than steel edgers.

Walking edgers have long handles that permit the finisher to stand while edging. Using them can save time on projects with many linear feet (meters) of formwork.

Groovers

Sometimes, groovers are used in conjunction with saw-cut joints to provide the depth required. The contract documents (plans and specifications) may require that the contraction joint be one-third the depth of the slab thickness. For example, a 6 in. (150 mm) thick slab may require a 2 in. (50 mm) deep contraction joint, while the plans call for tooled joints using a groover. In this case, the contraction joint can be formed with a groover the same day as the concrete placement. Next, the same joint formed by the groover can be saw-cut to the 2 in. (50 mm) depth once the concrete has taken final set sufficiently–typically 4 to 12 hours after concrete placement.

Hand Floats

Hand floating removes imperfections from earlier finishing operations, compacts the slab surface, and brings paste to the surface for troweling. Hand floats are often used before edging because screeding may drag concrete away from the form edge and leave high spots a few inches away. A hand float used in a fanning motion cuts any high spots and moves concrete to the edge so it's flush with the top of the form.

Hand floats are available in wood, magnesium, and composite materials and range in width from 3 to 4 in. (75 to 100 mm) and in length from 12 to 24 (300 to 610 mm). Wooden hand floats encourage proper workmanship and timing because they can stick, dig in, or tear the surface when they are used too early on any type of concrete. If wooden floats are used too late, they roll the coarser particles of fine aggregate out of the surface, at which time use of a magnesium float held in a flat position would be preferable. Wooden floats fill in low spots more easily with mortar. They should also be used in areas where embedded hardeners will be applied, floated, and finished by hand only. Wooden floats should not be used on concrete with a high air content because they tear the surface.

A hand float is used at the perimeter of the concrete fields to finish the edges flush with adjacent banding.

Magnesium hand floats require less effort. Like magnesium bull floats, they slide along largely on fines and work better on air-entrained concrete. They can be used on concrete from the time of placement to beyond the point of stiffening, when a wooden float cannot be used. Magnesium floats are best used in the initial smoothing of the surface near forms, walls, or other projections, and during placing, screeding, and bull floating, when a wooden float would dig in or tear the surface. Well-worn magnesium floats develop an edge almost as sharp as a steel trowel, so care should be exercised to use them flat to avoid closing the surface too early or causing blisters.

Composite hand floats using resin-impregnated canvas surfaces are smoother than wooden floats and only slightly rougher than magnesium floats. They are similar to magnesium hand floats and should be used in the same manner.

Finishers can use hand floats to achieve the final finish when a finely textured surface is desired for an outdoor slab, such as a sidewalk, patio, or driveway. Wood floats produce a very rough texture, but magnesium or resin floats produce a smoother texture that also improves slip resistance. A swirl pattern produced by an extra floating pass is another option.

Hand Trowel

Steel hand trowels are used to produce a smooth, hard, dense slab surface. No troweling should ever be done on a surface that has not been hand floated by either a magnesium or wooden float.

Hand trowels have open wooden handles and the blades are made of blue, stainless, or high-carbon steel. Sizes range from 3 to 5 in. (75 to 125 mm) wide by 10 to 24 in. (255 to 610 mm) long, with larger trowels being used for the first finishing pass, and a smaller trowel for each subsequent pass. The smaller trowel, sometimes referred to as a burn trowel, can be used when the concrete is in the final stage of setting.

Fresno Trowel

Fresno trowels are large trowels—2 to 4 ft (0.6 to 1.2 m) in length—that are used for final finishing after bull-floating and after all bleed water has evaporated. They can be used to rapidly trowel large areas and are fitted with extension poles so finishers don't have to walk on the slab. Fresno trowels produce a smooth, flat finish without the compaction produced when hand trowels of decreasing size are used with greater

A fresno trowel is used to further refine the concrete surface in preparation for surface retarder

edging of the trowel to put more pressure on the surface. This can be an advantage when troweling decorative work because the smooth finish doesn't seal the surface as much and is less likely to cause blistering or delamination.

Many ACI documents warn that hard-troweling air-entrained concrete can cause blisters and delaminations, especially if a dry-shake hardener is also applied. Much exterior decorative concrete, however, is placed in cold climates where air-entrainment is needed to make the surface resistant to damage caused by deicing salts and cycles of freezing and thawing. Some troweling is also needed to produce the best decorative results when dry-shake color hardener is used. Trowel only as much as is needed to produce the desired smooth surface and avoid sealing the surface by putting less pressure on the trowel.

Margin Trowels

These pointed mason's trowels can be used for a variety of purposes. When edging, it's a good idea to first use a margin trowel in an up-and-down cutting motion to move coarse aggregate particles away from the edge forms. That way, the side of the edger won't ride up on larger particles, dig into the concrete, or leave an uneven surface. Before grooving concrete with large coarse aggregate, use a margin trowel in an up-and-down cutting motion at the joint location. This also moves the coarse aggregate particles away from the joint location so the groover can be used without catching on the large aggregate.

Kneeboards

Also called sliders, these are flat pads that the finisher kneels on while hand-finishing concrete. Made of plywood, plastic, or aluminum, and padded with foam, they're essential not just for comfort but to also keep finishers' clothing from getting soaked by the caustic moisture in the concrete. Long-time exposure to moisture from fresh concrete can cause severe chemical burns to the skin.

Brooms and Brushes

Some decorative flatwork may require a slip-resistant finish. Driveways, sidewalks, pool decks, patios, and steps are examples. Finishers often use brooms, manufactured specifically for use with concrete, to create a slip-resistant surface while the concrete is still soft enough to achieve the desired depth of broom stroke.

Finishing brooms range in width from 24 to 48 in. (0.6 to 1.2 m). Handheld finishing brushes, good for texturing steps

A concrete finisher works with an edger and hand float to finish the perimeter of the concrete paving

and curbs, are 12 in. (300 mm) wide. Most brooms and brushes have plastic bristles, which are long-wearing and easy to clean. Blocks that hold the broom or brush bristles are made of wood or aluminum.

Bristle stiffness ranges from extra soft to stiff, depending on the desired finish. A coarse-textured broom finish is created with a stiff-bristle broom dragged across a floated surface. The surface has to be hard enough to prevent tearing or rolling of the finer aggregates. This brooming method removes the smooth texture created by edging or jointing tools. It is appropriate for handicap ramps.

A medium-textured broom finish is created by retooling the edges, floating the surface, and using a soft-bristle nylon broom to create the medium texture. If the pattern left by the edging and jointing tools is desired—as with a "windowpane" finish— the tools may have to be rerun along the joints and edges. The medium-textured broom finish is commonly used on driveways, patios, and steps, especially in areas exposed to rain, snow, or icy conditions.

A fine-textured broom finish is created by retooling the edges, floating the surface, and then making a troweling pass before texturing with a soft-bristle nylon broom. This texture may wear down over time and also provides less slip resistance than the other two broom finishes.

Saws

Although joints can be formed with a groover, sawed joints are often preferable because they are less noticeable. Dry-cutting methods are preferred because the water used for cooling wet-cut saws may affect color. Handheld saws or small walk-behind floor models can be used for dry cutting. The walk-behind

A saw equipped with a diamond blade makes decorative saw cuts using an aluminum straightedge

models ranging from 4 to 15 hp (3 to 11 kW) do a good job on surfaces imprinted with light to moderate textures, but are difficult to use on deeply textured patterns because the wheels don't ride level on the surface. Saw cut timing is important. Sawing too early kicks out pieces of aggregate at the edge of the joint (raveling). But, if the joint is cut too late, cracks may form at locations other than the joint. The waiting period before sawing may range from 4 hours in hot weather to 12 hours in cold weather. Sometimes the joints can be sawed as late as 24 hours after concrete placement if the concrete cannot be cut without spalling of edges.

Another option is an early-entry saw. These lightweight, dry-cut saws allow cutting much earlier (1 to 4 hours after finishing) than conventional saws because they use small-diameter diamond-impregnated blades and a skid plate that helps prevent raveling. However, the quality of the resulting joint should be reviewed and approved prior to using this method. Most models cut to a maximum depth of 1-1/4 in. (32 mm) but some larger saws can cut as deep as 4 in. (100 mm). The saw leaves a small windrow of dust at the edge of the cut joint that makes cleanup fairly easy. The major advantage of early-entry sawing is reduced likelihood of random cracking occurring before joints are cut. Cuts made with an early-entry saw can be chased at a later time with a traditional saw if a deeper or wider cut is desired.

Tools Used Specifically for Decorative Flatwork

With the exception of tools used to apply dry-shake color hardeners or release agents, most of the tools used specifically for decorative concrete are for imparting patterns, texture, or color to the concrete. Platform stamps made of aluminum, and later of plastic, were the first tools used to pattern concrete, but they added no texture and have largely been replaced by polyurethane stamping mats that impart both pattern and texture. There are many basic patterns such as brick, tile, slate, and cobblestone, plus more that mimic wood, fieldstone, and other stone patterns.

Stamping Mats

Use of rigid or semi-rigid stamping mats is dependent on the decorative finisher's preference. Rigid mats reportedly allow the stamping process to be started earlier, increasing the time window within which stamping can be performed before the concrete becomes too stiff. Semi-rigid mats are preferred by some finishers because they say they can better feel the consistency of the concrete while stamping. The semi-rigid mats also work better if the concrete surface is wavy. A stiffer mat might bridge over low areas without imparting as much pattern or texture.

Many stamping mats are sold in sets of several mats, each with a slightly different texture and variation in the pattern. This helps the stamper to avoid pattern repetition and produce more realistic looking results. Enough stamps are needed to cover the widest part of the concrete placement, plus two more mats to start the next row. As a simple example, with 24 x 24 (610 x 610 mm) stamping mats, you would need 10 mats to span a 20 ft (6 m) wide driveway, plus two additional mats. With irregular notched patterns, some stampers lay out more than a row at a time because it's more efficient. This requires having more stamps on hand. The more stamps you have on hand, the easier your concrete placement will be.

Stencils

Stencils are sheets of plastic or paper, with design cutouts. The stencil is laid flat on the concrete surface and color is added to produce a desired pattern. The first stencils used to create decorative concrete patterns were sheets of heavy-duty coated paper produced on a die-cutting machine and sold in rolls. The pattern was a running-bond brick. Finishers laid stencils on a floated concrete surface, embedded it with more float passes, and then broadcast a dry-shake color hardener uniformly over the surface. The dry-shake was either floated or troweled in. When the concrete had set enough to bear foot traffic, workers pulled up the paper pattern and applied a curing compound. Areas covered by the pattern remained uncolored and slightly indented, resembling mortar joints in the brick.

The process for use on fresh concrete surfaces, as opposed to overlays, has changed since stencils were first introduced. Many more standard patterns are available. In addition, a release agent can be applied before using a textured roller to produce a more authentic looking color and surface texture.

Stencils can also be used to produce patterns on overlays or on hardened concrete surfaces. These products may be made of paper or vinyl—with or without a sticky backing—or a thin, strong polyester film. When used to produce patterns on overlays, a grout-colored base coat is first sprayed or rolled on. After the base coat has dried, the stencil is applied and a stamp coat of a different color is troweled on. The stencils are removed when the stamp coat has set enough to be walked on.

Stencils used directly on hardened concrete surfaces are usually made of vinyl or thin polyester film. After the stencils are in place, the concrete can be acid-stained, dyed, sandblasted, or engraved to produce a pattern with or without texture. Some companies manufacture stencils in customized patterns sized to fit the concrete area. They can also be used to add logos or other personalized touches to a decorative concrete project.

Dry-Shake Spreaders

Dry-shake hardeners are usually broadcast by gloved hand, but sifting devices are available to permit good coverage near walls or in other hard-to-reach locations.

Tampico Brushes

Tampico brushes are commonly used to clean finishing tools but are also used to apply powdered release agents.

Flex Mats

Also called floppies, flex mats can be bought with the same texture and pattern as the stamping mats, but they are much more flexible. They bend easily and can be used to carry the pattern closer to walls or other obstructions. One or more may be needed for most stamping jobs.

Texturing Skins

Texturing skins are very thin and flexible, but impart only texture. They're commonly used along form edges and on vertical surfaces such as risers on decorative concrete steps. They can also be used to repair blemishes in texture caused by nonuniform pressure applied to stamping mats. Some finishers use them on larger areas when only texture is desired.

Tampers

Tampers are used to press the stamp firmly into the concrete surface and make a crisp imprint of the pattern, texture, or both. When using rigid or semi-rigid mats, metal dirt tampers available at most hardware stores work well. For flex mats or texturing skins, a polyurethane tamper with a wider tamping area can be bought from stamping tool suppliers. This tool won't cut the stamp, and the larger area also reduces the pressure so imprints of the tamper don't show up in the surface.

Pattern Rollers

Pattern rollers weigh approximately 55 lb (25 kg), are 36 in. (915 mm) wide, and imprint a number of different brick, cobblestone, flagstone, and slate patterns but without texture. An 8 in. (205 mm) wide roller produces brick pattern borders and a 16 in. (405 mm) wide roller produces other patterns for borders. A textured roller can also replicate some texture effects along with a pattern. The rollers are suitable for high production jobs but may not produce as realistic or aesthetically pleasing a result as mats because of the repetition pattern and texture. Small hand texture rollers can produce special finishes such as keystone or rock salt.

Hand Rollers and Chisels

Hand rollers are used after the stamping mats are removed to dress up grout joints that are part of some patterns. They can deepen joints when the mat wasn't tamped down with enough pressure and fix blurred pattern lines caused by stamps slightly overlapping a previously stamped surface. In patterns with very wide or deep joints, some paste or mortar may squeeze through the grout joint between stamps. This can be rolled out immediately with a hand roller. For patterns with narrow joints, however, hand rolling the still plastic concrete may widen the joint and make it look worse. In this case, chisels can be used after the concrete has hardened to remove the squeezed-up material. Chisels are also useful for enhancing pattern lines near side forms where the mat may have not made a full impression. All detailing work with rollers and most work with chisels should be done on the day of the placement.

Pump-Type Sprayers

Surface retarders, liquid release agents, colored curing compounds, and sealers can all be applied with a pump-type sprayer.

Concrete Tools

Steel trowels

Floats

Stair tools

Groovers

Grooved warning strip tools

Synthetic fibers

Tamper

Concrete tamp (sometimes referred to as a jitterbug)

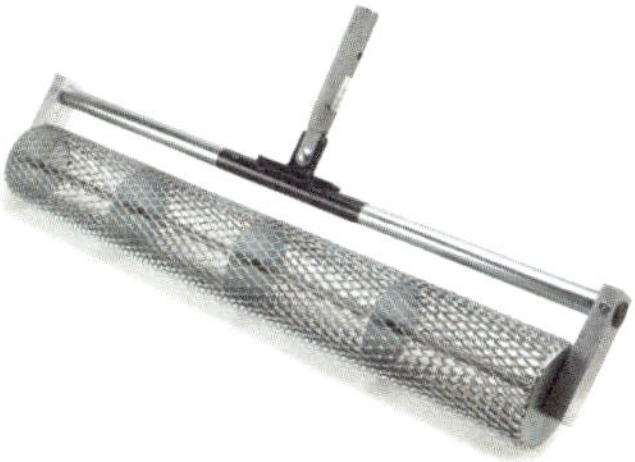

Rolling tamp

Small hand tamp

Large cutter

Small cutter

Funny trowel

Edgers

Dowel assemblies

Mason line

Mahogany float

Fiberglass float

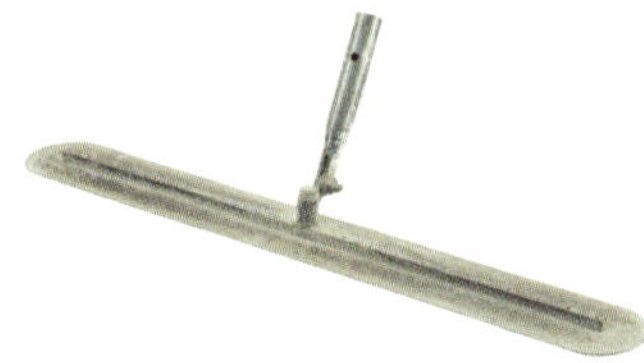

Fresno trowel

Metal stamp

Isolation joint material: spray glue or concrete nails can be used to attach the isolation joint material

Stamping tools

Finishing broom

Wooden kneeboard

Texturing skin – floppy tool

Pattern mat – floppy tool

Concrete sliders (kneeboards)

Finishing broom

Poly finishing broom

Texture roller

Narrow chisel

Wide chisel

Tampico brush

Margin trowel

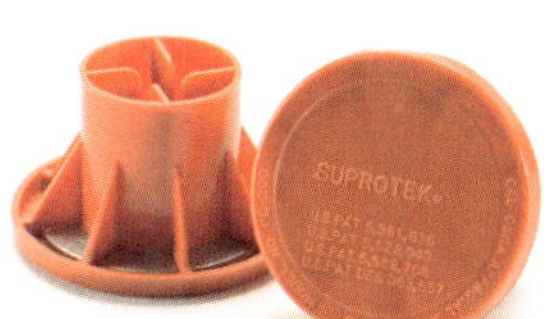

Reinforcing bar caps

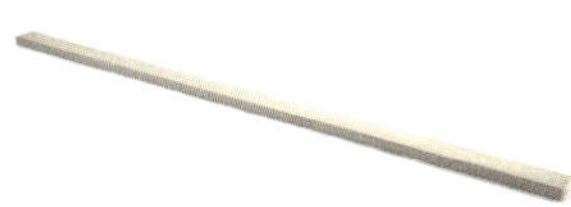

Aluminum straightedge

Aluminum extension pole

REFERENCES

1. ACI Committee 318, "Building Code Requirements for Structural Concrete (ACI 318-14) and Commentary," American Concrete Institute, Farmington Hills, MI, 2014, 519 pp.

2. ASTM C979/C979M-10, "Standard Specification for Pigments for Integrally Colored Concrete," ASTM International, West Conshohocken, PA, 2010, 5 pp.

CHAPTER 6: SURFACE RETARDERS

As stated in the section on "Surface Retarders and Set-Retarding Admixtures" in Chapter 5, surface retarders enable the achievement of light, medium, or deep surface exposures of the concrete's inner materials – primarily the sand, coarse aggregate, or a combination. A surface retarder in this regard is a thin film of liquid solution applied during the finishing stage while the concrete is still in its early plastic form (Fig. 6.1). It is not an integral admixture included or applied prior to placement.

Fig. 6.1—Surface retarders are pigmented to show coverage of the concrete

During the concrete's initial plastic state, retarding agents within the surface retarder's film layer diffuse into the concrete's surface paste to slow down the hydration reaction at the surface of the concrete. Some retarders have prescribed abilities to yield various etch depths. During the next 4 to 24 hours, the underlying concrete hardens except where the surface has been retarded. The concrete surface is then water washed to remove the retarded surface cement paste, revealing the underlying aggregate in the concrete.

General Uses

Concrete surface retarders are used to expose the sand or coarse aggregate at the concrete surface. Exposure levels can range from the sand for a "micro" (fine) sand-wash finish up to 1-½ in. (38 mm) nominal maximum size aggregate for a "macro" (very coarse) type finish.

A variety of decorative concrete applications may use surface retarders to achieve exposed surface finishes, including walkways, driveways, patios, pool decks, plaza decks, stairs, curbs, and walls, as well as in combination with other surface texturing techniques. Surface retarders can be used as an aid by applying them prior to sandblasting to reduce the time and damaging effect of the blasting to both the cement matrix and the aggregate used, particularly when specialty materials or decorative color are specified.

Surface retarders are available in many grades or "etches" to produce multiple finishes of varying degrees with the same concrete mixture, depending on the manufacturer. The varying grades of surface retarder create differing textures and appearances.

Top-Surface Retarders

Top-surface retarders are generally comprised of water-based, low-viscosity liquids that are sprayed on the finished surface of freshly placed concrete after initial set has occurred. They differ from in-form, paint-type surface retarders by the level and type of ingredients. They also have the ability to be applied at varying thicknesses, yet maintain the required degree of penetration and uniformity of outcome.

Top-surface retarders ordinarily must be protected by plastic covering to remain active and to protect them from premature etch loss or crusting of the concrete surface. The plastic covering should not be placed on top of the surface retarder until it has cured to a fine film. This typically occurs 1 hour after applying the surface retarder, depending on the weather. In warmer weather, the surface retarder can be washed off the same day that it is applied if the concrete has gained sufficient strength.

The concrete matrix is allowed to harden normally while the surface is controlled by the surface retarder. The surface retarder inhibits the reactions that take place normally over 4 to 24 hours during the initial stages of concrete strength gain. When the bulk of the concrete has sufficiently hardened, the surface mortar can be washed with water and a broom, pressure water-washed, or sandblasted away, exposing the underlying aggregate (Fig. 6.2).

Fig. 6.2—This fine, nonslip finish for use at an aquatics water park was created by using a surface retarder

Guidelines for Use

Preparation

In preparation for using a surface retarder, consider the following:

- Determine the manufacturer and select the desired level of light, medium, or deep surface exposure for the concrete surface finish.
- Design a mixture that combines the materials, including the integral color if required, you want to see on the final surface of the concrete. Special aggregates can also be broadcast on the surface of the concrete and floated into the concrete per the project specifications. Refer to Chapter 8: "Seeding Special Aggregates." Seeded aggregates less than 3/8 in. (10 mm) diameter are not recommended for use with a surface retarder, as washout may occur and they may not adhere well over time. Consult with your local ready mixed concrete producer for the available aggregates, integral colors (if desired), or cement types available, as well as the appropriate concrete mixture to achieve the desired performance and finish outcome.
    - Light exposure surfaces may require that a greater proportion of the total aggregate in the mixture be sand (for example, 50% sand or greater), and deep exposure surfaces may require that a greater proportion be coarse aggregate (for example, between 60 and 70%).
- Ensure that you have both enough personnel and proper quality finishing tools to perform the finishing work required that will meet final expectations.
    - Review the project's placing and finishing techniques, and related application step timings, before commencing actual placing and finishing.
    - Review the surface retarder removal and containment plans before proceeding with placing and finishing.
- Confirm that a sufficient amount of surface retarder solution is available onsite to support the project's required surface area to be exposed, and in accordance the manufacturer's instructions.
- Always construct a mockup sample for the client or design professional's review and approval. Mockup samples are beneficial to the concrete contractor and

the client. They define the expectations of the finished product and this should not be discounted.

Applying the Surface Retarder

Following are the steps used to apply a surface retarder:

- Rod or screed the placed concrete in a sawing motion to remove excess material and leave the aggregate at the surface.
- Float sufficiently to bring cement paste to the surface, enabling proper closure of the surface.
- When producing micro-sand finishes, the use of a rolling tamper is most effective in developing a uniform mortar content that can be troweled smoother, thereby creating a tighter, denser sand finish. The use of a rolling tamper will allow the use of concrete mixtures containing less sand and more coarse aggregate. This may be necessary to produce the required water-cementitious materials ratio or meet a specified compressive strength, for example. The concrete mixture proportions must conform to the requirements of the soils report, structural drawings, or other project plans or specifications.

A saw cut is placed between two integral color concrete panels for cleanliness in design

- Avoid larger aggregate showing on the perimeter of micro-exposed (fine sand) surfaces. A small hand tamp can be used to slightly depress the coarse aggregate adjacent to the concrete forms prior to finishing the concrete. After edging the first time, floats should be used to swipe paste back over the edge of the form. This is done by placing the float onto the concrete surface approximately 6 to 10 in. (150 to 250 mm) out from the form at a 45-degree angle to the form, and swiping back to the edge. The float should be kept flat for this procedure.
- When producing a micro-sand finish, generally two or even three additional passes with steel fresno trowels or

hand trowels are necessary to ensure a dense, smooth, ridge-free surface from which the sand and additional fines can be exposed.

- Avoid kneeboard marks showing in the final surface. Only non-absorbent materials should be used. Absorbent materials such as wood can leave dark spots on the concrete. If impressions are left in fresh concrete by kneeboards, they should be removed prior to final finishing and before applying the surface retarder.
- If bleed water persists, wait until it dissipates prior to application, but be sure to apply the surface retarder when the concrete still has sufficient moisture remaining.
- In accordance with the manufacturer's recommended application coverage, apply the top surface retarder soon after the final finishing operation, using a low-pressure or garden-type sprayer (note: follow the manufacturer's canister and spray tip recommendations).
- Cover with plastic after the retarder has crusted over. Otherwise, staining from the plastic sheeting could occur. This will help achieve uniform results without crusting on the surface, allowing for easier removal of the retarded cement paste the following day, depending on the conditions and products used.
- Some surface retarder products may not require plastic covering. Consult the manufacturer's guidelines and recommendations and perform test panels prior to commencement of work whenever using surface retarders.

Surface retarder is washed off of the concrete surface after the concrete has sufficiently hardened

Removing the Surface Retarder

Remove the surface retarder in accordance with the manufacturer's instructions. In especially hot weather, it may be necessary to remove the retarded material the same day. Late removal may result in difficult removal and affect the consistency of the final finish.

The surface material removed includes retarded hydrated cement paste, fine aggregate, and coarse aggregate that can result in alkali wash water and cement solids. This material should be handled carefully and disposed of according to local regulations. Sand dams, diversionary dikes, or a vacuum with sufficient storage capacity should be used to prevent alkali runoff during the removal and cleaning process.

Collect all solid particles and allow water to evaporate and dispose of the solid particles as normal construction waste.

Environmental Note:

Surface retarders are used to achieve a uniform sand finish or to expose larger aggregates in a concrete mixture without using acids or sandblasting. Both acid washing and sandblasting are environmentally unfriendly and pose defined health risks. Surface retarders produce a similar result without the use of caustic chemicals, dust, noise, or gasoline or diesel engines.

It is required to divert and collect all residues and washoff during the finishing process to eliminate potential environmental contamination. Refer to "Removing the Surface Retarder" for specific guidelines in this regard.

Removal timing depends on the specific surface retarder used, the environmental conditions and temperatures, and the mixture proportions. Removal could be as early as 2 to 3 hours and as late as 24 hours after applying the surface retarder. The key is that the underlying concrete must be hard enough so it is not damaged by whatever means is used for washing.

When the treated area has cured (usually overnight), high-pressure washing is the most effective means of removing the retarded material. If high-pressure washing is not available, street brooms or stiff bristle brooms and a garden hose can yield satisfactory results. Depending on the application, a commercial floor scrubber with a nylon bristle brush in combination with a water hose may be used for removal and should be available on the project when washing begins.

Tools Required

Surface-retarded concrete requires the use of normal finishing practice tools, unless otherwise indicated herein. These include:

- Water supply
- Water hose
- Water pressure washer (1500 to 2500 psi [10.3 to 17.2 MPa])
- Standard low-pressure pump sprayer similar to the type used to spray curing agents
    - Sufficient quantity of sprayers should be on hand to support the required application rate and provide backup
- Sprayer tips for application coverage in accordance with manufacturer's recommendations
- Stiff bristle broom. Don't use metallic brush brooms on decorative concrete.
- Plastic sheeting to protect all adjacent surfaces and structures from overspray
- Materials to contain the runoff. Do not allow the washoff to be discharged to storm drain systems.

Curing and Sealing

To enhance the beauty and color of the aggregates and protect the surface, the use of a breathable acrylic sealer or a penetrating siloxane or silane sealer is recommended to protect the finished surface from damage. This is particularly important where salt damage from inclement weather is a concern.

A surface retarder finish can be cured with a moisture-retaining cover that does not cause discoloration. Curing compounds and wet curing methods are not options due to the application of the surface retarder. The curing method selected should be performed during the construction of the mockup sample and prior to commencement of the actual work.

Durability and Sustainability

Surface retarders create a concrete surface texture that is nonslip and requires limited maintenance. The concrete may use gray cement, white cement, or integral color; all three have different light reflectance values.

Light-colored concrete, with its high surface relectivity, reduces surface temperatures. This results in cooler pavements and roofs, where concrete could be used as a topping slab. This produces a reduced heat island effect. In addition, light-colored concrete could reduce the amount of lighting required due to its reflectance values.

Refer to the surface retarder manufacturers' MSDS information for safety requirements, and review them with field personnel prior to using. When working with surface retarders, avoid contact with skin and use eye and respiratory protection.

CHAPTER 7: STAMPED CONCRETE

Fig. 7.1—Stamped concrete design achieved through the use of texturing tools

Stamped concrete is a cast-in-place, monolithic slab that is colored and imprinted with a pattern, texture, or both (Fig. 7.1). Properly specified and installed, stamped concrete can resemble many materials such as cut stone, random stone, slate, brick, and tile. The tools can also create three-dimensional texture.

Stamped concrete is usually installed with color hardener or integral color and often uses contrasting release agent if a textured surface is required. The release agent remains in the low points of the texture to create color variations and highlights.

Stamped concrete can be placed indoors and outdoors and is often used for pool decks, walkways, plazas, patios, driveways, and steps.

Planning and Organization

In typical summer weather conditions, it takes about 3 to 5 hours to place, finish, and stamp concrete. During this time, at least 10 different steps must be completed, compared with four steps for exterior broom-finished concrete. Under hot weather conditions, the time available can be as little as 2 hours. When ready mixed concrete temperatures exceed 85°F (29°C), care and pre-planning are necessary to have an organized crew on hand to finish the work. It is best to place stamped concrete in the morning hours to reduce the problems with an accelerated set time.

In cold weather, 6 to 8 hours can pass before initial set. Contractors need accelerating admixtures to manage the placement and avoid excessive bleeding and poor impressions. If three workers are needed to place 1000 ft^2 (90 m^2) of plain gray concrete in a day, plan on five or more to place and stamp 1000 ft^2 (90 m^2) of stamped concrete. Some contractors will not exceed 100 to 125 ft^2 (9.3 to 11.6 m^2) per worker.

Subgrade Preparation

Subgrade preparation is critical for any slab-on-ground construction, but additional care should be exercised with decorative concrete applications. The subgrade must be uniformly graded and compacted throughout. This helps ensure that the concrete will be of uniform thickness. It is recommended that the thickness of the fill match the thickness of the slab.

Well-drained, compacted fill, such as a 2 in. (50 mm) thick sand base, allows the concrete to set uniformly and cure with a more consistent color. There should be no soft spots in the subgrade, otherwise the concrete must bridge the soft spots, increasing the possibility of cracking. Before placing the concrete, make sure there is no standing water or muddy, frozen, or spongy spots. During hot-weather placement, the subbase may be moistened in advance, as long as the base is dry to the touch at time of placement. Excess moisture in the subgrade can also carry additional salts to the surface during hydration and curing. The additional salts can cause an increase in efflorescence, resulting in discoloration or spotting.

Concrete Mixture Proportions and Admixtures

The concrete mixture proportions should meet any and all requirements for a specific purpose and geographic area. For exterior paving, the concrete mixture could range from 2500 psi (17 MPa) for pedestrian paving in a mild (non-freezing) climate to 4500 psi (31 MPa) for a climate with frequent cycles of freezing and thawing. A typical integral color stamped concrete mixture will require 5-½ to 6-½ sacks of cement per cubic yard (307 to 362 kg/m^3). For concrete that will be stamped, a "fatty" or "rich" mixture is recommended; one with enough cement paste to provide a layer of cream on the surface that will take an imprint well. A 4 to 5 in. (100 to 125 mm) slump is recommended.

For impressions not deeper than ¼ in. (6 mm), a ¾ to 1 in. (19 to 25 mm) nominal maximum aggregate size works well. For deeper impressions you may need to go to a 3/8 to ½ in. (9.5 to 13 mm) nominal maximum aggregate size using aluminum or "cookie-cutter" tools.

Pozzolans and other supplementary cementitious materials (such as fly ash and slag cement) can have positive effects on concrete such as improved finishability, reduced permeability, reduced efflorescence, and minimized bleeding of color.

Admixtures can also be helpful. Retarding admixtures can slow the set of concrete in hot weather, buying additional stamping time. In cold weather, an accelerating admixture will speed setting time and minimize bleeding and segregation. Avoid using admixtures containing calcium chloride, as these produce additional efflorescence and can have an adverse effect on the final color.

Most water-reducing and retarding admixtures are compatible with color hardener applications. To improve durability without sacrificing workability, a water-reducing admixture may be used. These are designed to lower water requirements by approximately 10%, improving strength while reducing permeability and cracking.

Formwork

Sloppy forming can make stamping more difficult and can devalue the finished product. It is imperative that the formwork be properly squared to any adjacent structures or buildings. For jobs requiring repeating square or rectangular stamp patterns (such as brick or tile), the forms must be perfectly square at the corners. Most patterned texturing tools will fit together in a squared pattern, showing any variance if the formwork is not squared. Diagonal braces keep forms in alignment when the weight of fresh concrete pushes against them.

It is also critical to take precise elevation measurements. Proper slope must be established to ensure adequate drainage. The recommended slope for exterior paving is a minimum 2%. However, adhere to the ADA Guidelines as required. Maximum cross slope for a sidewalk is 2%, thus no tolerance.

Be sure to drive or cut form stakes flush or slightly below the top edge of the forms so as not to interfere with stamp mats.

Coloring Stamped Concrete

There are two primary ways to color concrete: 1) broadcasting color hardeners onto the surface of freshly placed, bull-floated concrete; and 2) mixing integral color into ready mixed concrete. Powdered or liquid release agents can be used to provide a highlighted or mottled appearance.

Color hardeners (made up of portland cement, iron oxide pigments, and finely graded silica sand) are used to color the surface of fresh concrete applications (Fig. 7.2). It is not a topping, but actually becomes part of the concrete surface. Color hardener also increases surface density and abrasion resistance, providing increased durability. These steps should be followed:

Fig. 7.2—Color hardener is worked into the concrete surface with a wood bullfloat; note the bright color.

- After placement and screeding, initial floating should be done on air-entrained concrete using a wood bull float. This allows bleed water to reach the surface more easily where it can evaporate. The wood float also gives the surface a good profile for accepting the color hardener.
- After allowing the bleed water to dissipate from the surface, the color hardener is broadcast onto the surface by hand or using a spreader. A particle mask or respirator should always be worn during this operation, as the color hardener contains silica. The color hardener should be applied at the specified manufacturer's rate for the product. Typical applications will use approximately 60 lb per 100 ft^2 (290 kg per 100 m^2) of surface area. Lighter color hardeners, such as off-white or cream colors, may require 100 lb per 100 ft^2 (49 kg per 100 m^2) or 1 lb per 1 ft^2. The first coat should take two-thirds of the material required for the area. This should be allowed to "wet out," or absorb available moisture in the concrete, before attempting to float the surface.
- The color hardener should be floated using either a wood or magnesium bull float depending on the site conditions, ambient temperature, and exposure to wind or sunlight. A steel hand trowel or fresno trowel should never be used during the color hardener application, as this could result in delamination.
- After floating the first application of color hardener, the remainder of the material should be applied to the surface. Apply the second coat at a 90-degree angle to the first application.
- A steel hand trowel may be used before texturing if a smoother surface is required, but is not generally necessary. During hot weather applications, an evaporation retarder or monomolecular film may be used on the surface to help reduce rapid moisture loss and crusting. Always apply the film after the hardener application and not before, as the film may prevent the concrete moisture from wetting out the hardener, resulting in delamination. A surface-applied water-reducing admixture may also be used to help reduce surface crusting.
- Integral color, which comes in powdered, granular, or liquid form, is typically mixed into the concrete at the batch plant, creating uniform, consistent color. An advantage to using integral color is the labor and time savings of not having to float the color in or wait for it to absorb moisture. It can be used to color vertical as well as horizontal surfaces. Hues, however, are more subtle, and less vibrant than those from color hardeners.

Release Agents

A release agent must be applied as a bond breaker before the texturing process begins, preventing the tool from sticking to the plastic concrete. Two types are commonly used: one a powdered form and the other a liquid, hydrocarbon solvent. The powdered material is typically tinted and used as an accent or highlight color. The liquid material can be clear or pigmented. The release agent should be applied right before the texturing process begins and after the color hardener application (if applicable) and final troweling have been completed. The concrete should still be in a plastic state when the release agent is applied.

Powdered release agent should be fluffed by gloved hand or with a paddle to aerate the material before use. Broadcast the material to the surface by hand or by using a 6 in. (150 mm) mason's brush at the specified manufacturer's rate; typically approximately 3.5 lb per 100 ft^2 (17 kg per 100 m^2). The material should be distributed evenly over the surface and also used to condition the texturing tools before use. Do not allow the material to form clumps or thick areas on the surface, as blank spots in the texture may result. Not using enough powder can cause concrete to stick to the stamp, ruining the finish. Never trowel or impact the release agent into the surface in any manner. The surface may only be touched with the texturing tools after application.

While powdered release agents provide greater contrast in color than liquids, greater care needs to be taken, as on windy days the particles can become airborne and stain objects.

Liquid release agent is applied using a low-pressure pump sprayer following the manufacturer's coverage rate. In hot weather, never spray more than one or two rows ahead of the texturing crew. The material may evaporate before the texturing tools reach the area. Liquid release agent does not require wash down, as the material will completely evaporate from the surface. Liquid release can be followed up with colored antiquing agents to provide finish color.

Stamping/Texturing Tools

Texturing tools are available in many patterns representing natural materials, including slate, stone, and wood, and other man-made materials such as brick, cobblestone, and tile. Tools today are typically made from a polyurethane material and include a pattern and a texture. Some tools are designed as pattern-only tools, like the old-style aluminum cookie cutter tools. Many are still in use today, but the polyurethane tools are the most versatile and popular.

To determine if the surface is ready to accept a texture requires experience and familiarity with the concrete mixture, the ambient conditions, the texturing tools, and the coloring system. The concrete should be firm enough to support the weight of the person applying the texture, but still plastic enough on the surface to accept the texture. The surface should also be void of any bleed water or residual moisture produced by troweling.

The surface resistance may be determined by applying pressure with your thumb. The surface should offer resistance yet still leave a clean impression of 3/16 to ¼ in. (5 to 6 mm) while not exposing any aggregate. It takes experience and skill to know when the surface is ready to accept texture. The timing is critical; too early will cause fuzzy texture or birdbaths, while too late may result in little or no texture. The texturing application should always follow the same direction as the concrete placement.

Seamless Texturing Tools

Seamless texturing tools are a simple and effective way to texture large areas and areas of random size and shape. These tools produce no specific pattern that is apparent on the surface, only non-directional texture. This type of texture also allows the joints to function as part of the design.

Three or four tools are sufficient for most areas, depending on the size of the tool and the size of the area receiving texture. A combination of 48 in. (1.2 m) tools (or larger), with one or two smaller ones for edge detail, is the most versatile combination. Typically, the amount of tools required will be enough to cover the width of the placement, plus one more tool returning in the opposite direction.

After the release agent has been applied, the first tool may be set in place. Care should be taken not to drag the tool across the surface, as this will cause undesired blemishes and imperfections. The next tools should be placed allowing an overlap of 4 to 6 in. (100 to 150 mm). The tools may then be set using a pounder or tamping tool. Be sure to set the tools evenly and avoid using the pounder on the edges as they may be set hard, leaving an outline of the tool in the surface. After setting the first tool, reverse the overlap before moving it to the next position. Lift the edge of the second tool, then lift the first tool edge to the top position. That way, when the first tool is lifted it will not drag on the concrete, causing defects in the surface. Rotate the tools often during placement to randomize the surface texture. Use the smaller tools for texturing the edges and wall lines or any place that the larger tools do not easily fit. Continue to move the tools in this manner until the entire area has received texture to reduce repeating the pattern.

Patterned Texturing Tools

Patterned texturing tools produce a visible geometric pattern on the surface as well as the texture associated with the pattern. Careful planning is required to be sure that the pattern will stay square to existing structures. Planning for proper joint placement should also be considered in the pattern choice and layout. It is good practice to fit the tools together in advance to be certain they fit properly and that the technician understands the proper sequence for row alignment and random placement.

For pattern texturing, enough tools are required to span the area in one direction and begin the next row. For example: if the area is 12 x 24 ft (3.6 x 7.2 m) and the texturing tools are 24 in. (0.6 m) tools, you would need six tools to span 12 ft (3.6 m) plus two more to begin the next row—a minimum of eight tools.

Once the concrete is ready to accept texture and has been properly released, the first tool may be placed. When possible, the first tool should be placed at a square edge or against a structure to ensure that the tools will remain square during the application. This is not necessary with random pattern tools. Placement of the first tool is critical, as all other tools will be set in relationship to the first. If the first tool is out of square, the entire area will reflect this.

Continue placing the tools in the proper sequence to complete the first row. All tools may be placed at this time. Each manufacturer will have a system of color coding, numbers, or a combination of letters and numbers to help randomize the texture on the surface. The tools should always be placed in a manner suited to this purpose.

Keep the tools fitted tightly together. Do not slide or move a tool after the initial placement, as defects and double impressions may occur.

Set the tools evenly using a pounder. Never pound the leading edge of a tool unless a backer tool has been placed first. This will help prevent blowout damage caused by trapped air escaping from the center of the tool.

Moving in the same direction that the concrete was placed, continue to move the tools forward. Keep the rows tight together until the entire area has been imprinted. As the tools are moved forward, the joints and edges should be detailed using detail tools that match the texturing tools. These may include rollers, strikers, chisels, and small touch-up skins.

Stamped concrete pavement

Curing the Concrete

The curing method chosen is critical for textured concrete durability and the overall quality of the project. Refer to the copy of "Curing Decorative Concrete, Problems & Practice #4,"[1] by the Decorative Concrete Council reproduced in Appendix A. The main purposes for curing concrete are to minimize cracking, increase durability, and reduce the effects of early carbonation on the surface. Use of color hardener will increase the surface durability and abrasion resistance significantly. In addition:

- The curing method should not damage the surface or the color of the fresh concrete, but should commence as soon as possible after finishing is completed.
- Use of an evaporation retarder or monomolecular film can help prevent rapid moisture loss from the surface of the concrete during the plastic state.
- Physical membranes may be used for curing as long as they are breathable and leave no marks or differential curing patterns on the surface. Never use plastic sheeting. The best way to cure stamped concrete–with release agent–is by using a curing cover lined with a woven fabric. The curing cover goes on after the concrete stamping process has ended and the concrete has sufficiently hardened. The curing cover is placed directly over the release agent; water is not placed on top of the release agent prior to covering.
- The curing method should be demonstrated in the mockup panel for test purposes.

The need for adequate curing is more critical under extreme weather conditions. If the weather is hot or windy, covering the concrete with a layer of impervious building or kraft paper can help prevent rapid moisture loss. Early morning placements are advised in hot weather. It is recommended to avoid placing decorative concrete during windy conditions. In cool weather, the concrete should be covered with insulating blankets, especially if the overnight temperatures may drop below freezing. Blankets help retain the moisture and heat needed for proper curing.

Early Carbonation

Early carbonation results in a soft, dusty concrete surface. It is most prevalent when concrete is placed in cold weather and is in the presence of a source of carbon dioxide gas such as unvented heaters. The presence of carbon dioxide gas is also very dangerous to the workers. Carbonation is the chemical combining of carbon dioxide with the hydration products of portland cement. Carbon dioxide combines principally with calcium hydroxide, but will also attack and decompose the calcium silicate and aluminates and combine with the calcium portion of these compounds to form calcium carbonate. Concrete is most susceptible while it is in its plastic state. As temperatures dip below 40°F (4°C), proper hydration of concrete is severely slowed. The concrete stays in the plastic state for a longer period of time, and if a source of carbon dioxide is present, the likelihood of carbonation increases. Carbonation affects concrete to various degrees from a light dust on the surface to a depth of 1/4 in. (6 mm) or more.

To lessen or prevent carbonation damage to floors placed during cold weather:

- Use an accelerating admixture in the concrete.
- Use temporary heaters that do not allow combustion gases to contaminate the atmosphere surrounding the paving at an interior concrete placement.
- Open the building to fresh air during interior concrete placement to ventilate exhaust fumes from trucks, troweling machines, and other equipment.
- Use a membrane curing and sealing compound as soon as possible to block the carbon dioxide gas from the concrete. A uniform and heavy film will be necessary. It has been proven that burlap and polyethylene sheets for curing do not offer adequate protection. The concrete is most susceptible in the first 24 hours.

Wash Down

Textured concrete that has been treated with a powdered release agent must have the release agent removed before the project can be sealed. This is usually done 24 to 48 hours after the texturing has been completed, depending on the ambient temperature, concrete mixture, curing method, and the manufacturer's recommendations. The concrete must be strong enough to withstand the cleanup.

Removing the excess release agent will create the highlighted look often associated with textured concrete. Care should be taken not to remove too much of the material, but enough should be removed to prevent any problems with the sealer application. Typically 70 to 80% of the release agent is washed away. The amount of material that will be removed should be demonstrated on the mockup panel.

The method used for removal should be determined by the manufacturer's recommendations and the site conditions. Although power washing is a common method of removal, it can cause damage to fresh concrete surfaces if done too early. Using a low-pressure hose with a combination of cleaners and brushes is a safer, less aggressive method. Stubborn patches of release agent may require more aggressive spot scrubbing. Any material removed from the surface should be collected and properly disposed of in accordance with local regulations.

Sealing

Colored, textured concrete should always be sealed to protect the surface from UV deterioration and provide additional chemical and wear resistance. Certain types of sealers may darken or enhance the color of the surface. Most solvent-based sealers result in a glossy sheen. Products that provide a low-gloss or matte finish are also available. The sealer chosen for the project should be applied to the mockup panel to show the finished result.

Sealing should not be confused with curing, as each has its own purpose and are generally done at different times. Sealing should never be done before the concrete has been allowed to hydrate for a period of time, allowing excess moisture to leave the slab. Sealers must be applied to a clean, dry surface. The sealer manufacturer should give specific recommendations for the time required.

If efflorescence has appeared on the surface, it should be removed before sealing. This may be done with a cleaner or a very light solution of muriatic acid diluted with 20 parts of water. The area should be pre-wetted, scrubbed lightly with the cleaner or solution, and rinsed completely. If an acidic solution or cleaner is used, it may be necessary to neutralize the surface before sealing. The surface should be allowed to dry for 24 to 48 hours before applying any seal coats. Always follow the manufacturer's recommendations for the specific product used.

REFERENCES

1. Decorative Concrete Council, "Curing Decorative Concrete: Problems & Practice #4," American Society of Concrete Contractors, St. Louis, MO, Mar. 2006, 1 p.

CHAPTER 8: SEEDING SPECIAL AGGREGATES

Surface retarders can be used to expose the sand and aggregate contained within the concrete. However, they are frequently used to expose special aggregates that are broadcast and worked into the concrete surface (Fig. 8.1). The advantage of using special aggregates is obvious. Natural stone, specially crushed granite, and other crushed or round gravels and stones come in a variety of colors. The stone, in conjunction with or without integral color, produces a hard, durable, attractive concrete paving.

Seeded aggregate works equally well for vehicular or pedestrian paving. It is nonslip and hides stains well due to the surface irregularity. A darker integral color and aggregate works well in vehicular paving (motor courts, entry drives, and parking lots), as it serves to hide oil and tire marks.

Lighter aggregate colors, such as off-white or cream-colored aggregates, may help contribute to LEED points, as they increase the albedo (light reflectance) of the concrete.

The aggregate selected should satisfy the requirements of ASTM C33/C33M, "Standard Specification for Concrete Aggregates,"[1] or ASTM C330/C330, "Standard Specification for Lightweight Aggregates for Structural Concrete."[2] The aggregate selected should be sound.[3] Soft aggregates[4], reactive aggregates[5-7] (glass, shells, and pyrite, for example), and sedimentary stones (slate, for example) should be avoided, as they may wear or pop out of the concrete over time. Alkali silica reaction (ASR) causes an expansive silicate gel to form with reactive aggregates present in concrete. Reactive aggregates include glass, flint, chert, and certain rhyolites. The expansive force of the silicate gel is greater than the tensile strength of concrete, so the concrete will crack. Depending on the reactivity of the aggregate and the alkali content of the cement, cracking can occur as early as

Fig. 8.1—Special seeded aggregates exposed using a surface retarder

days, or it can be delayed for many months or even years. As cracking occurs, more water enters the interior of the concrete to mobilize the alkalis present. In severe cases, ASR cracking can destroy the strength of the concrete and ultimately reduce it to rubble over the years.

Aggregates that contain organic impurities[8] or degrade under abrasion[9] should be avoided.

Typically, aggregates ranging from 3/8 to 1 in. (9.5 to 25 mm) in diameter are broadcast onto the surface of the concrete at a rate of 2 to 4 lb/ft^2 (10 to 20 kg/m^2). Aggregates of this size may be exposed using a surface retarder. Finer aggregates, such as silica carbide and feldspar, are typically exposed using muriatic acid or light sandblasting. However, sandblasting will dull the aggregates and is more likely to dislodge aggregates from the concrete surface.

The aggregates are broadcast onto the surface before the concrete sets and must be worked into the concrete. Knowledge of the specific gravity[10,11] or density[12] of the aggregate may be useful because low specific gravity aggregates can float on the mixture or high specific gravity aggregates can sink too deeply into the concrete to be exposed. The cement paste, also known as cream, is brought up through the aggregate using a float.

Once the aggregates are coated in paste, the surface is sealed with a steel trowel. A funny trowel works well to reach out and seal the surface. The surface is troweled until it is flat, smooth, and free of any trowel marks or surface blemishes.

The appropriate surface retarder, selected for the size of the aggregate, is applied to the concrete surface at the manufacturer's dosage rate. A 100% coverage with the surface retarder is required.

Depending on the weather and the concrete characteristics, the concrete can be washed either the same day or the next day. If the concrete is to be washed to expose aggregate the next day, then the surface should be covered with plastic sheeting to contain moisture.

A mockup panel using the selected aggregate, dosage rate of aggregate (in lb/ft^2 [kg/m^2]), and depth of exposure should always be completed prior to the commencement of work.

REFERENCES

1. ASTM C33/C33M-13, "Standard Specification for Concrete Aggregates," ASTM International, West Conshohocken, PA, 2013, 11 pp.
2. ASTM C330/C330M-14, "Standard Specification for Lightweight Aggregates for Structural Concrete," ASTM International, West Conshohocken, PA, 2014, 4 pp.
3. ASTM C88-13, "Standard Test Method for Soundness of Aggregates by Use of Sodium Sulfate or Magnesium Sulfate," ASTM International, West Conshohocken, PA, 2013, 5 pp.
4. ASTM C142/C142M-10, "Standard Test Method for Clay Lumps and Friable Particles in Aggregates," ASTM International, West Conshohocken, PA, 2010, 2 pp.
5. ASTM C1260-14, "Standard Test Method for Potential Alkali Reactivity of Aggregates (Mortar-Bar Method)," ASTM International, West Conshohocken, PA, 2014, 5 pp.
6. ASTM C1567-13, "Standard Test Method for Determining the Potential Alkali-Silica Reactivity of Combinations of Cementitious Materials and Aggregate (Accelerated Mortar-Bar Method)," ASTM International, West Conshohocken, PA, 2013, 6 pp.
7. ASTM C1293-08b, "Standard Test Method for Determination of Length Change of Concrete Due to Alkali-Silica Reaction," ASTM International, West Conshohocken, PA, 2008, 7 pp.
8. ASTM C40/C40M-11, "Standard Test Method for Organic Impurities in Fine Aggregates for Concrete," ASTM International, West Conshohocken, PA, 2011, 2 pp.
9. ASTM C131/C131M-14, "Standard Test Method for Resistance to Degradation of Small-Size Coarse Aggregate by Abrasion and Impact in the Los Angeles Machine," ASTM International, West Conshohocken, PA, 2014, 5 pp.
10. ASTM C127-15, "Standard Test Method for Relative Density (Specific Gravity) and Absorption of Coarse Aggregate," ASTM International, West Conshohocken, PA, 2015, 5 pp.

11. ASTM C128-15, "Standard Test Method for Relative Density (Specific Gravity) and Absorption of Fine Aggregate," ASTM International, West Conshohocken, PA, 2015, 6 pp.

12. ASTM C567/C567M-14, "Standard Test Method for Determining Density of Structural Lightweight Concrete," ASTM International, West Conshohocken, PA, 2014, 4 pp.

CHAPTER 9: DECORATIVE TREATMENT OF STAIRS

It can be challenging to achieve the color and texture required by project plans and specifications for cast-in-place concrete stairs, while at the same time conforming to building codes. Cast-in-place concrete stairs serve as a transition between surfaces of differing elevations (Fig. 9.1), and good planning is important. There are two components of stairs: treads are the horizontal portion and risers are the vertical portion.

Fig. 9.1—Stairs with integral color and a sand finish achieved through a surface retarder

Code Requirements

The ADA Guidelines[1] contain the following requirements for stairs:

- On any given flight of stairs, all steps are required to have uniform riser heights and uniform tread widths. Stair treads are required to be no less than 11 in. (280 mm) wide, measured from riser to riser. Risers are not permitted to exceed 7 in. (177 mm) in height (Fig. 9.2). Open risers are not permitted.
- The undersides of nosings are not permitted to be abrupt. The radius of curvature at the leading edge of the tread is not allowed to be greater than ½ in. (13 mm) (Fig. 9.3). Risers are required to be sloped or the underside of the nosing is required to have an angle not less than 60 degrees from the horizontal. Nosings are not allowed to

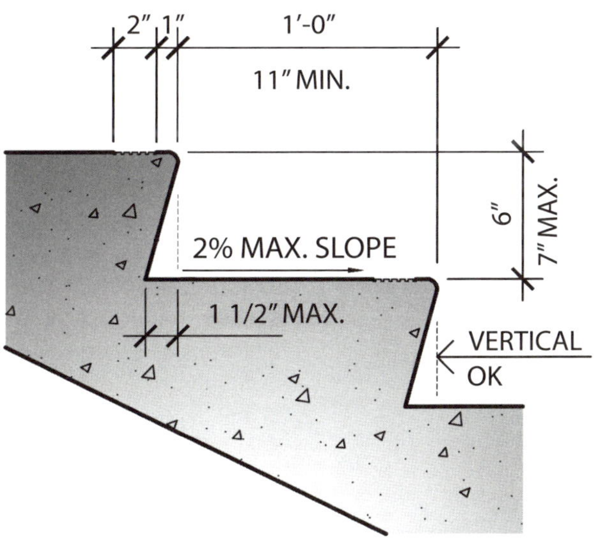

Fig. 9.2—ADA Guidelines include: 11 in. (280 mm) minimum treads, 7 in. (177 mm) maximum risers, a nonslip surface, a grooved strip with contrasting color, and ½ in. (12.5 mm) maximum radius nosing (Note: 1 in. = 25.4 mm)

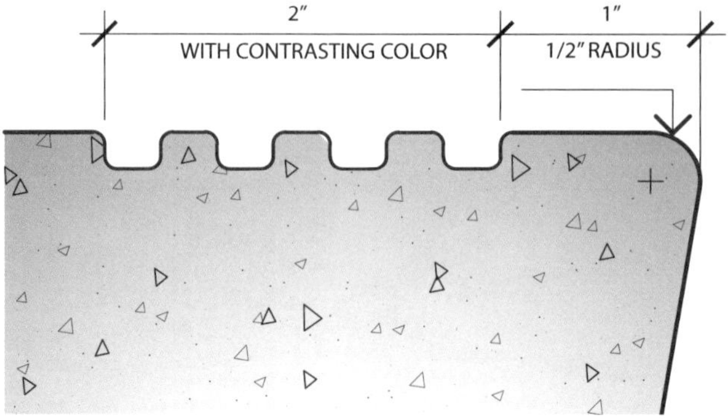

Fig. 9.3—Tooled warning strip

project more than 1.5 in. (38 mm) (Fig. 9.2).

- Outdoor stairs and their approaches are required to be designed so that water will not accumulate on walking surfaces.

The 2012 International Building Code[2] (2012 IBC) allows a 7 in. (197 mm) maximum riser height, except in special circumstances involving utility and miscellaneous structures (Group U). Typically, the allowable range for risers is from 4 to 7 in. (100 to 178 mm) in height. On any given flight of stairs, all steps must have uniform riser heights and uniform tread widths.

For buildings in the United States, stair geometry—especially riser height—is tightly controlled and enforced. Section 7.2.2.3.5 of the National Fire Protection Association's Life Safety Code, 2012 Edition,[3] requires riser height to be measured as the vertical distance between tread nosings. When the tread slopes, this height can be measured using a carpenter's level and tape measure, as illustrated in Fig. 9.2.

Section 7.2.2.3.6 of the Life Safety Code requires that there be no variation in excess of 3/16 in. (5 mm) in the depth of adjacent risers, and the tolerance between the largest and smallest riser shall not exceed 3/8 in. (10 mm) in any flight. ACI 117-10, "Specification for Tolerances for Concrete Construction and Materials (ACI 117-10) and Commentary,"[4] contains the same 3/16 in. (5 mm) tolerance for the depth of adjacent risers and 3/8 in. (10 mm) requirement for the difference between the maximum and minimum riser depth.

Concrete stairs must be constructed with great care in both the forming and finishing phases as corrective action is difficult and can be very expensive. One reason corrective action is expensive is that modification of one tread elevation will affect two riser heights. That is, grinding or patching the surface of one tread may bring a riser into tolerance, but it can simultaneously cause the adjacent riser to be out of the specified tolerance. In a cascading effect, each successive adjacent tread may require grinding or patching until the bottom or top of the flight is reached. Often, grinding or patching will result in poor appearance, and although the steps may conform to code, they may be rejected for aesthetic reasons.

The 2012 International Residential Code[5] (IRC) R311.7 requires that concrete steps must be a minimum of 36 in. (914 mm) wide. Treads must be a minimum of 10 in. (254 mm) in width, measured from riser to riser (Fig. 9.2).

Stairs are typically constructed before the adjacent concrete flatwork is placed to them; therefore, the location and the elevation of the top and bottom of the stairway must be correct. Commercial projects will typically have a grading plan that will call out the changes in elevation, as well as the number of risers required. A riser approximately 6 in. (150 mm) in height, and a tread approximately 12 in. (300 mm) in width (measured from riser to riser) is a practical starting point if a stair design is not given in the project documents. Maintain consistency in riser height and tread width at all given stair locations to conform to code requirements.

Color and Texture of Cast-in-Place Concrete Stairs

Decorative concrete stairs may be finished in the following ways:

- Stamped
- Broomed
- Exposed aggregate or sand finish using surface retarder
- Other specified specialty finishes, such as sandblasting or seeded aggregate

Each type of finish will require a different application process to produce the desired result. The areas that will remain unchanged are layout, determining elevations and calculating riser height and tread width, basic formwork, setting slope for tread drainage, bracing, and placement.

Required Tools (Specialty Tools Required)

Basic finishing and carpentry tools are required for setup and placement. Additional specialty tools required may include:

- Stair nosing tool
- Stair inside tool (cove tool)
- Stamping tools
- Specialty forms (plastic)
- Specialty layout tools (available from various manufacturers)
- Small pump type sprayer for liquid release agent (if required) or surface retarder

Formwork

Formwork is the single most important aspect when installing stairs and risers. The forms must be capable of

Fig. 9.4—Concrete steps reinforced with vertical stakes as well as stringers placed above the top of the riser forms

Fig. 9.5—Bevel on a riser board to allow finishing of the tread

Fig. 9.6—Riser faces can be battered or vertical; layout is placed on the edge form

containing the concrete in its fresh state, while allowing the finisher access to surfaces that require finishing.

With traditional forming (Fig. 9.4), the bottom edge of the riser should be beveled to allow a hand float to reach the face of the step (Fig. 9.5).

The forms for the risers (Fig. 9.6) must be easy to remove after the initial set so the finisher can apply the required finish to the face of the riser. Several specialty tools and forms are available if a contractor decides to simplify the forming process. Any formwork, traditional (Fig. 9.7) or otherwise, should be treated with a form release agent to permit the forms to release cleanly from the concrete. The form release agent selected should be verified to have no effect on the concrete color.

Finishing Practices

The maximum slump for use in the placement of stairs and risers should not exceed 4 in. (100 mm). The placement of the concrete requires proper consolidation to ensure there are minimal air pockets or voids near the face. If a vibrator is used, do not over vibrate because it will force concrete down into the lower steps.

The riser form must be stripped wet so that the faces may be finished. The timing of this operation will vary with the ambient conditions and the set time of the concrete. The concrete should take an initial set but still be able to accept a finish. It is good practice to remove one edge of the top riser and watch to see if the face slumps or dips slightly; if it does, the riser form should be refastened until additional time has passed.

Fig. 9.7—Forming a large set of stairs on grade takes precise layout

Procedure

Concrete is typically placed at the top of the set of steps and placement continues down to the bottom step. This makes it easier to place the concrete from one step down to the next. Place the concrete evenly and consistently. Do not let concrete build up behind one portion of a riser, as the forms may move.

Forms should be tapped lightly to release air bubbles that could cause bugholes.

Float the surface of the first (top) step until it is relatively smooth and at the proper elevation. Note that all excess concrete should be placed in the step below. Continue filling and floating the steps, working downward until all the steps have been filled and floated.

After initial set of the concrete, the steps will be finished from the top step down to the bottom step. Place kneeboards on the third and fourth step from the top if the concrete is hard enough.

Kneel on the top kneeboard. Float the top step until it is relatively smooth and at the proper elevation. Use a level to ensure that the tread has the proper fall. Treads should not exceed 1/4 in. (6 mm) fall from the back of the tread to the front of the tread. Hold the trowel relatively flat against the concrete surface and trowel the step.

Edge the nosing, along the riser form, with the specified tool. This may be an edger with a 1/4 to 1/2 in. (6 to 13 mm) radius (maximum) or a tooled warning strip tool, if applicable. Continue troweling the treads and edging the nose until all steps have been completed for the first time.

After all of the step treads have been troweled the first time, there is normally a waiting period until the steps set sufficiently to hold their shape when riser boards are removed. At the proper time, each riser board should be carefully removed.

Generally, the risers will be finished with a float to fill in any voids and smooth out the vertical surface. The inside step tool is moved back and forth along the cove until the concrete is smooth and uniform in appearance. The outside step tool, manufactured with the same radius as the edger used previously, is moved back and forth along the top edge of the riser. Kneeboards are moved down one step at a time until all steps have been completed. Addi-

Fig. 9.8—Top surface retarder applied to stairs placed on a waterproofing membrane over a structural slab

tional finishing steps depend on color method and texture required.

In a wider (more than 5 ft [1.5 m] wide) set of steps, two finishers can work side by side to complete the task more efficiently. Timing is important! The key is to work cleanly, carefully, and quickly because too much time on any one step may cause others to set too firmly for proper finishing.

It is recommended that a mockup of the cast-in-place stairs be completed to gain approval and clarify any issues with color, texture, or constructibility.

Integral Color

Placing and finishing steps often includes the use of integral color. This method works well with all types of finishes, including stamped concrete, exposed aggregate, sandblast finishes, and broom finish. Always follow the manufacturer's recommendations for mixing, placing, and finishing to ensure consistent color. Note that vertical faces will tend to appear darker. This is due to diffuse reflections associated with the vertical surface, as opposed to a horizontal surface that receives direct light. Water-cementitious materials ratio must be maintained when using integral color. The slump should also be consistent. Typically, a low-slump concrete should be placed for stairs and risers. The average consistency should not exceed a 4 in. (100 mm) slump.

Stamped Treads and Risers

When placing and finishing stamped concrete steps, it is also good practice to keep the consistency at or below a 4 in. (100 mm) slump. The finishing process will depend somewhat on the coloring method.

The simplest method of coloring stamped steps is to use integral color or no color. Another method of coloring that requires more preparation is the use of a color hardener. A third method of coloring is the use of a release agent (Fig. 9.8). These come in a powder or a liquid. The powdered form offers the addition of a highlight color while the liquid form is a clear hydrocarbon solvent. If using color hardener, all of the horizontal surfaces (treads) should receive color hardener, in two separate shakes, and be floated before removing the riser form. Some color hardener may be mixed with water to a thick consistency, to use on the vertical faces. After determining the proper timing, the riser forms should be removed and the faces should be floated to fill any voids. The color hardener mixture may be applied using floats (rubber, sponge, fiberglass, or magnesium). The first coat of "plaster" serves as a scratch or bond coat. An additional thin coat of the color hardener mixture may be applied using a trowel or float immediately following the scratch coat. The top and bottom edges should be blended with the stair treads at this time using the step radius tools or other specialty finishing tools. After the entire surface has been properly troweled, the release agent can be applied. Follow the manufacturer's recommendations for proper application. The texturing tools should also be conditioned with the release agent. Texture the areas as quickly as possible using caution not to tamp too hard near an edge (tamping by hand or by using a hand float is the best method), so you do not break or damage the top edge of a face. After the entire area has received texture, the area can be detailed using chisels and hand tools to impart joints or simulated stones. The concrete should be allowed to set for the recommended period of time before removing the release agent. Refer to the manufacturer's recommendation for sealing.

Exposed Aggregate or Sand Finish

With exposed aggregate or sand finishes, the typical coloring method is integral color, or no color. However, color hardener may be used to achieve a sand finish as well.

Consult the surface retarder technical data sheet, provided by the manufacturer, for specific application details. A surface retarder should be selected that will provide the desired depth of aggregate exposure.

After floating and troweling all of the surfaces, the surface retarder is applied using a garden type sprayer once the concrete has been completely finished. The surface retarder is applied to the risers and the treads until 100% coverage is achieved. Depending on the time of year and the retarding product being used, the time to wait before washing may vary from a few hours to the next day. If the stairs are to be left overnight it is good practice to cover them completely with plastic sheeting. This will hold moisture in place at the surface and prevent the retarder from being accidentally removed. Washing the steps and risers should follow the same procedure as the flatwork. The main difference is that a nylon brush may be necessary for the riser faces. Two complete rinses should be completed to remove excess cement paste. A final wash and cleaning should be completed before sealing.

Contraction Joints

Saw-cut contraction joints may be placed in stairs to reduce the appearance of shrinkage cracks during the curing process. Contraction joints should be placed on stairs wider than 8 ft (2.4 m) or as the concrete mixture, soil conditions, or environmental conditions require. Stairs should typically be saw-cut within 24 hours of the stair placement.

The joints are laid out on the top of the treads and the face of the risers. A diamond blade is used to cut and weaken the concrete to a depth of approximately 1 in. (25 mm). The intersection of the tread and riser, commonly called the stair cove, is difficult to saw-cut and a smaller hand grinder in combination with a smaller radius diamond blade can be used to access this area.

Building Stairs with a Stamped Texture

The steps for constructing stairs with a stamped texture are illustrated in Fig. 9.9.

Summary of Stamped Steps

1. Concrete is placed from the top riser down to the bottom riser using a maximum 4 in. (100 mm) slump concrete with 1 in. (25 mm) maximum size coarse aggregate. Mixtures containing smaller aggregate may require starting from the bottom riser and working up.
2. The treads are floated and the nosing, where the tread meets the riser, is tooled with an edger. A ½ in. (13 mm) radius is typical.
3. The riser forms are stripped and the treads and risers are floated to a consistent texture.
4. Color hardener is floated onto the treads and risers, and then troweled to a smooth finish.
5. Release agent is broadcast liberally to the treads and risers.
6. Textured concrete mats are stamped into the surface of the concrete to provide texture to the treads and risers.
7. Chiseled joints are hammered into the treads and risers, if required.
8. The concrete is cleaned and sealed after a 7-day cure.

Cast-in-Place Steps with a Surface Retarder

The steps for constructing stairs with a surface retarder are illustrated in Fig. 9.10.

1 - Stripping riser forms

2 - Floating tread and riser to a consistent texture

3 - Applying the color hardener in a paste form

4 - Floating tread and riser to a consistent texture

5 - Color hardener has been troweled, ready for release agent

6 - Application of contrasting release agent

7 - Stamping of treads

8 - Stamping of risers

9 - Layout for chisel work

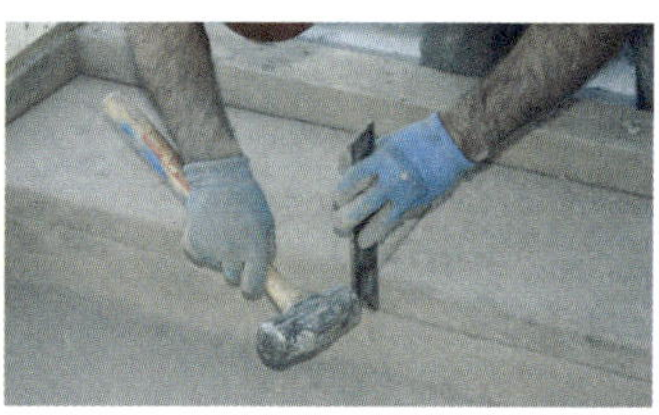

10 - Chisel work is applied to treads and risers to achieve the design.

11 - Washing off release agent with a pressure washer

Fig. 9.9—Steps for producing stamped concrete stairs

1 - Finishing of concrete treads

2 - Finishing of concrete risers

3 - Final touches on grooved tread warning strip

4 - Application of surface retarder

5 - Treads and risers with surface retarder applied

6 - Washing off the surface retarder with a pressure washer to reveal the aggregate

7 - Consistent sand texture on the risers and treads

Fig. 9.10—Steps for producing concrete stairs with a surface retarder finish

REFERENCES

1. *2010 ADA Standards for Accessible Design*, Department of Justice, Washington, DC, Sept. 15, 2010, 275 pp.

2. *International Building Code 2012*, International Code Council, Falls Church, VA, 2011, 692 pp.

3. *NFPA 101: Life Safety Code 2012*, National Fire Protection Association, Quincy, MA, 2012, 489 pp.

4. ACI Committee 117, "Specification for Tolerances for Concrete Construction and Materials (ACI 117-10) and Commentary," American Concrete Institute, Farmington Hills, MI, 2010, 76 pp.

5. *International Residential Code 2012*, International Code Council, Falls Church, VA, 2011, 904 pp.

CHAPTER 10: SEALING DECORATIVE CONCRETE

Properly sealed and maintained concrete will better weather the test of time

Sealers, and the problems associated with them, continue to be the one of the largest sources of frustration for decorative concrete applicators. It's easy to understand why. There are a vast number of sealer products on the market for decorative concrete, yet many contractors lack a basic understanding of how to select and properly apply the products. This section will address the most common sealer problems encountered in the field, why they occur, and how to avoid and repair them.

While most sealers straight out of the container look, smell, and flow similarly, there are major differences. Here's a brief overview of the four types of sealer products used for decorative concrete. All have a different function, purpose, and recommended method of application.

Curing Compounds

Cures, also called curing compounds, are designed to slow the evaporation of water from the concrete surface, allowing for more complete hydration of the surface paste. However, they are not intended to provide long-term protection. They are applied as soon as concrete finishing is complete. Caution must be taken to verify color matching with colored concrete.

Sealers

Sealers provide long-term protection and color enhancement. They should not be applied until after concrete has cured. The recommended minimum curing time is 28 days, but most contractors typically wait only 7 to 14 days. Table 10.1 provides a comparison of different types of concrete sealers.

Cure and Seals

Cure and seals, as you might expect, combine some of the benefits of cures and sealers. Like cures, they slow the evaporation of water from the concrete surface, allowing for more complete hydration of the surface paste. They also provide midterm protection of 6 to 12 months and color enhancement. These products are applied as soon as concrete finishing is complete.

Coatings

Coatings provide long-term protection, the best chemical resistance, and color enhancement. Like sealers, they must be applied after the concrete has fully cured (28 days). They may also require special surface preparation for proper adhesion.

Note that cures, cure and seals, and straight sealers all have a dry film thickness of approximately 1 mil (25 μm) and are breathable. Coatings are thicker (2 to 3 mil [50 to 75 μm]) and typically nonbreathable. Even with so many options and chemistries, the problems with these products, such as environmental impact and application issues, are similar no matter what product is used.

VOC Regulations — How They Affect the Products You Use

Volatile organic compounds (VOCs) are emitted as gases from certain solids or liquids, including many types of sealers. VOCs include a variety of chemicals, some of which may have short- and long-term adverse health effects. In addition to sealers, VOCs are emitted by a wide array of products, including paints and lacquers, paint strippers, cleaning supplies, pesticides, building materials, and even furnishings.

Depending on the environmental regulations in effect at the project location, VOC limits range from 50 to 400 grams/liter (g/L). The typical VOC content in solvent- and water-based concrete cures, cure and seals, and sealers ranges from 50 to 780 g/L. It is your responsibility to use a product that meets VOC guidelines. Consult your regional Environmental Protection Agency (EPA) office or local government environmental office to obtain the VOC limits in your area.

Not all solvents contain VOCs. The government uses a complex formula to determine which solvents are harmful. As local, state, and federal governments clamp down on VOC emissions, manufacturers have been forced to reformulate their products to bring them within VOC compliance (or eliminate them from the market altogether). This means that the formulation of sealers and coatings you are accustomed to applying may have changed. The most common change has been a reduction in xylene (a VOC producer) and an increase in acetone (a non-VOC producer). While this switch has virtually no effect on the protection afforded by the sealer or coating, it can have other consequences, such as

Table 10.1—Comparison of Concrete Sealers (courtesy of Concrete Network [Anne Balogh])

Sealer Type	How They Work	Primary Applications*	Type of Finish	Performance
Penetrating Sealers (Including silanes, siloxanes, and silicates)	Penetrate and react chemically within the capillaries of the concrete to shield against moisture penetration and deicing chemicals.	Exterior concrete surfaces subject to corrosion and cycles of freezing and thawing. Where a natural, matte finish is desired.	Provide invisible protection without changing the surface appearance or leaving a sheen.	Provide excellent protection against outdoor exposure conditions. Most products are also breathable, allowing moisture vapor to escape.
Acrylics	Form a thin protective film on the concrete surface. Available in both solvent- and water-based formulations.	Both exterior and interior concrete. On projects where easy application and economy is important. To enhance the beauty of colored, stamped, or exposed-aggregate concrete. On fast-track projects, because acrylics often dry to the touch within an hour.	Available in a range of sheen levels, solvent-based acrylics generally enhance color better than water-based products.	Provide good protection against water and chloride intrusion, but usually wear faster than polyurethanes and epoxies. Solvent-based acrylics generally perform better than water-based products for outdoor use. On indoor surfaces, softer acrylic sealers usually require regular maintenance with several coats of a sacrificial floor finish, or wax, to prevent wear and black marks.
Polyurethanes	Form a high-build protective film on the concrete surface. Available in both solvent- and water-based formulations.	Both exterior and interior concrete. On floors in high-traffic areas, to provide good resistance to scuffs and staining. To enhance the beauty of colored, stamped, or exposed-aggregate concrete. Concrete countertops.	Available in a range of sheen levels. Finish is transparent and non-yellowing.	Nearly twice as thick as acrylic sealers, and produce a very durable chemical- and abrasion-resistant finish. Most urethanes are moisture intolerant until they cure, so no water should be present on the surface when the sealer is applied.

*Note: Always check with the sealer manufacturer to verify the compatibility of their product with the decorative surface you plan to apply it to.

making the product difficult to roll out, faster drying times, and an increase in spider webbing and surface blistering.

Preventing Moisture Problems

Moisture is a leading cause of problems with decorative concrete sealers. Under certain conditions, moisture can become trapped in or under the sealer, resulting in whitening or clouding of the sealer membrane. Why does this happen, how do we avoid it, and how do we fix it?

There are two key contributors to moisture problems. The first is sealer contact with moisture in the concrete during application. Cures, cure and seals, and sealers for decorative concrete are all designed to handle different levels of moisture contact. Cures and cure and seals can handle higher levels of moisture contact, allowing them to be applied to green (high moisture content) concrete without turning white or cloudy. Decorative concrete sealers, on the other hand, can't take much moisture contact. This is why they need to be applied after the concrete has cured for 28 days. If a decorative sealer is applied to green or wet concrete, it is very likely to develop an ugly white haze. This has to do with the type of resin (or plastic) the coating is made from and how that resin deals with moisture contact.

The second key contributor to moisture problems is the permeability of the sealer, or how readily water is able to pass through the sealer membrane. Permeability is directly related to the solids type and content and thickness of the sealer. All exterior acrylic cures, cure and seals, and sealers are designed to allow some level of permeability when applied at 300 to 500 ft^2/gal. (7.4 to 12.3 m^2/L). The lower the solids content or the thinner the membrane thickness, the more moisture can pass through the sealer without getting trapped and turning white. This is why applying sealer at the proper thickness is so important, especially when dealing with high-solids-content products (in excess of 25% solids). The higher the solids content, the smaller the margin of error. Most moisture-related problems in the field are caused by over-application of high-solids cure and seals or sealers.

Avoiding moisture-related problems is quite simple. Use a sealer with a solids content of less than 25%, and apply a thin coat by spray (Fig. 10.1). In a worst-case scenario, it may be necessary to strip off the sealer, followed by cleaning of the surface and sealer reapplication.

The Effects of Deicing Salts on Sealer Performance

Deicing salts have no direct effect on sealers. Sealers for decorative concrete often fail in areas where deicing salts are applied, or that receive drip-off from parked cars. It's not the salt, but rather what the salt is doing, that causes the sealer to fail.

Salt chemically reduces the temperature at which water freezes. When salt is applied to a sealed decorative concrete surface covered by snow and ice, it causes melting and turns the frozen water into a liquid that is now able to migrate into the concrete. This salt-rich water (brine) goes through many cycles of freezing and thawing as environmental conditions change. In some areas, it's possible to have hundreds of freezing-and-thawing cycles a year. During each cycle, the water expands as it freezes and contracts as it thaws. The problem is that while sealers help retard moisture movement, they do not stop it completely. So, as the saltwater passes through, under, and all around the sealer, the water is expanding and contracting, and eventually the sealer will fail (Fig. 10.2).

In areas with severe winters, some contractors use a combination of sealers to fight the effects of deicing salts. They start with a penetrating sealer (silane, siloxane, or silicone) that fills the concrete pores from the bottom up. Then, they apply an acrylic sealer that creates a membrane from the top down. This systems approach costs a bit more, but when faced with stripping and resealing, it may be well worth it.

Fig. 10.1—Sealing a patio wall and steps with an airless sprayer

Fig. 10.2—Sealer failure caused by deicing salts often starts out with a white, hazy appearance, similar to what you see when moisture is trapped by the sealer. Actually, a very thin air gap has formed between the sealer and concrete due to water pressure. Eventually, the sealer will pop off the surface.

The Effects of Temperature on Sealer Reactivity

The second major cause of problems (after moisture) when applying sealers to decorative concrete is temperature. Both air and surface temperature play a role, but surface temperature is typically more critical. After application, sealers undergo a chemical reaction that causes them to cure and form a film. Temperature plays a critical role in how fast or even whether this reaction occurs. The best temperature range for applying sealers is 50 to 90°F (10 to 32°C). That 40°F (22°C) window is really not very big, especially when you're working outside. Monitoring weather conditions should be mandatory before every sealer application. The next section will discuss what can happen if temperatures are too low or too high.

Low Temperature

Every sealer has a minimum film forming temperature (MFFT), or the minimum temperature needed for the sealer to properly form its film, cure, and get hard. For most sealers, this temperature is around 40 to 45°F (4 to 7°C) or higher. To be safe, most sealer manufacturers specify 50°F (10°C) to provide a buffer zone. If the temperature is at or slightly below the MFFT, the chemistry of the sealer is affected, the reaction slows down, and there is partial to no film development. Bottom line: the sealer is weak and will not hold up very long. If the temperature is really cold, film development stops altogether and the result is a white powder on the surface after the carrier (solvent or water) evaporates.

High Temperature

Practically every chemical reaction is affected by temperature. In the case of a sealer, as the temperature increases, so will the reactivity of the sealer. Increased reactivity decreases the working time, or pot life, of the sealer. The faster the reactivity, the less time the sealer has to wet out the surface, de-gas, and form its film. This makes it critical to get the sealer down on the concrete quickly and efficiently. As temperature increases, the ability to roll out sealers becomes more difficult. It is generally recommended to spray solvent-based sealers, especially in warm conditions. A common indication that the temperature is too high is the formation of fine "spider webs" or "cotton candy" strings coming off the roller or spray tip. This occurs when higher temperatures cause the solvent to flash (evaporate rapidly due to pressure change) before the resin (plastic) in the sealer can form its film.

Another common issue caused by higher temperatures is the formation of bubbles or blisters in the sealer. They occur when the solvent evaporates too quickly, trapping gas and air in the sealer. With today's tightening VOC requirements, more fast-evaporating solvents are being used, which reduces the window of application even further. When outside temperatures are expected to rise above the recommended application range, apply the sealer during the cooler times of the day—typically mornings and evenings.

Dealing with Condensation on Sealers

What happens when moisture and temperature both come into play? The air that surrounds us always contains water vapor, but the amount of water can vary. Humidity is the measure of how much water is in the air at any given time. We would not have to worry about this water vapor if it just remained trapped in the air as a gas. But, it doesn't because temperature fluctuations convert that water vapor into a liquid. If temperatures rise and enough water is in the air, instability is created and rain can fall. As temperatures fall, condensation can occur in the form of dew. For example, on cool summer nights, you'll often see dew-covered cars, grass, and other surfaces once morning comes around. The dew point is the temperature at which water comes out of the air and becomes a liquid.

As humidity increases and temperature decreases, water will condense on cool surfaces. Because concrete is like a sponge, it absorbs the condensation. The problem is that the slab surface won't look wet, but hiding just beneath it can be lots of collected water. If a sealer is then applied to the surface, the trapped water can cause the sealer to white out or not adhere properly. Outdoors, this problem is more common during transition seasons (spring and fall) as nights get colder but humidity is still high during warm days. Indoors, this problem is prevalent in the winter near walls and doors where floor temperatures are colder.

Preparing Surfaces for Sealer Application

A simple but often overlooked step in any sealer application is surface profile. Surface profile, in regard to sealing, includes all aspects of the surface at the time of sealer application. The two major factors are cleanliness and porosity. Overlooking either can cause even the best sealers to fail.

Clean means just that: clean! A surface that is to be sealed must be free of all dirt, dust, and any other contaminants that will come between the sealer and the surface. Spending a little extra time cleaning can make all the difference in how well the sealer adheres. In some cases, a good broom or blower is all that is needed to remove loose dirt. More stubborn contaminants may require removal by scrubbing with soap and water followed by a clean water rinse or acid etching followed by neutralization. Residue from stains, dyes, excess release powder, and efflorescence are also surface contaminants. These types of dry contamination are most often the culprits when sealers fail due to a dirty surface. Efflorescence and stain residue are especially nasty because their extreme pH levels can affect sealer chemistry. A sealer that exhibits white "curds" in the film or soft spots is often failing due to a surface pH imbalance.

Porosity refers to the concrete surface's ability to absorb the sealer. If the sealer can't wet out, there will be little or no adhesion and durability. A hand-troweled concrete surface is usually porous enough to accept a one-part sealer with a solids content lower than 30%. A machine-troweled concrete surface will usually require additional prep to open it enough to accept the same sealer. Typical methods for opening a very tight or dense surface include light sanding or acid etching. When dealing with higher-solids sealers (usually two-part polyurethanes and epoxies with solids in excess of 45%), opening the surface or diluting the first sealer coat is highly recommended. A simple water test (to see how well the water wets out the surface) is a great way to determine if the surface is ready to accept the sealer.

As with any sealer, always refer to the manufacturer's installation guidelines for specifics on surface preparation and proper application techniques.

Choosing the Best Applicator

How a sealer is applied can affect the final outcome and performance as much as all the environmental factors. Using the right tools is critical to achieving the best coverage rate and sealer thickness for optimal performance. While there are some general rules or guidelines for applying sealer, always refer to the manufacturer's installation instructions.

The most important rule is that less is more. Apply sealer in multiple thin coats rather than one thick heavy coat. To get this liquid sealer down on the surface in the most efficient manner—and give it the best chance to harden and not fail—ask yourself the following questions:

- Is the sealer solvent- or water-based? The liquid component of the sealer will determine how fast it evaporates.
- What is the solids content (usually stated in percent solids)?
- What is the resin type—acrylic, polyurethane, or epoxy?
- The answers to these questions will help in the selection of the best applicator type.

High-volume, low-pressure (HVLP) sprayer or airless sprayers: These sprayers are usually the best way to apply any sealer. They permit very controlled application rates, allowing large areas to be sealed in the shortest time. They will also handle both water- and solvent-based sealers, high and low solids contents, and any type of resin. The downside is the cost. You may want to consider renting because this spray equipment can be expensive.

Pump-up or low-pressure sprayers (Fig. 10.3 and 10.4): These are best for applying one-part, solvent-based sealers with a solids content below 35%. Make sure to use a sprayer that can handle solvents, and clean out the sprayer with clean solvent to avoid clogging the lines.

Roller: A paint-type roller with a 1/4 to 3/8 in. (6 to 10 mm) nap can be used to apply both water- and solvent-based sealers. However, when the solids content is high (above 35%), rolling becomes a problem for solvent-based sealers. Rough surfaces and high temperatures can also present problems when rolling solvent-based sealers due to puddling and surface bubbling. Consider spraying followed by back rolling when sealing stamped or textured surfaces with a solvent-based sealer.

Lamb's wool applicator: This is basically a wool rag wrapped around a wooden block. It's good for applying high-solids (above 35%) solvent-based sealers, especially polyurethanes and epoxies. But its use is limited to smooth surfaces because the applicator will not run over a rough surface.

Microfiber applicator: This synthetic-fiber pad is best for applying water-based, low-solids sealers on smooth surfaces and is a great way to achieve very thin coverage. Push and pull the sealer around, maintaining a wet edge.

Synthetic closed-loop mop: Use this applicator for water-based finish sealers (waxes) only. Move the finish around with the mop, maintaining a wet edge.

Fig. 10.3—Small pump-up sprayer

Fig. 10.4—Large pump-up sprayer

Sealer Application Tips

Each type of sealer has a recommended applicator and coverage rate, as discussed in the section "Choosing the Best Applicator." But simply using the right application tool won't guarantee good results. You also need to practice the proper application techniques to avoid bubbles, blisters, lap lines, and other imperfections.

The most common application problem is applying too much sealer at once. Sealers are designed to perform best at a specific thickness, depending on the type of resin. This is determined by the coverage rate for a particular sealer. The first and second thin coats are very stable, have good adhesion, and provide good durability. The more you apply, either in one or multiple applications, the more unstable the system becomes. With solvent-based systems, the signs of over-application are typically bubbles, blisters, and white haze. With water-based systems, blisters, foam, or a milky white cloudiness often form.

Another common application mistake is lap lines, or uneven application. When applying sealer, go back over the previous pass approximately 2 in. (50 mm) as you move across the surface. This overlap needs to occur when the sealer is still wet, so the two passes blend and become one. If the first pass dries, the second creates a lap line and can be seen after the entire floor is dry. Fixing the problem usually requires applying another full coat of sealer.

When applying sealer by sprayer (whether HVLP, airless, or pump-up type) make sure to maintain constant pressure and use the proper tip. A cone-shaped spray pattern is better than a fan pattern. The more atomized the sealer, the better.

When applying sealer by roller, make sure to buy a roller suitable for the sealer type (water- or solvent-based) and a nap thickness appropriate for the surface. When rolling on water-based sealers, be careful not to over-roll, which can cause foaming and blisters. You also may need to dip the roller more often. Some newer acetone-based fast-drying sealers can't be roll applied because they evaporate too quickly.

When using a lamb's wool applicator, microfiber applicator, or synthetic mop on smooth (typically interior) surfaces, the application process is the same. Pour the sealer on the surface, and push and pull the product while maintaining a wet edge until you achieve the desired thickness. These application methods are very good for water-based sealers because they don't foam and you can see the white sealer go clear as you push and pull it around the floor. However, they will only work on smooth floors.

REFERENCES

1. Sullivan, C., Concrete Network, www.concretenetwork.com. (last accessed 2012)

CHAPTER 11: COLOR AND TEXTURE SAMPLES

Seeded Aggregate Finishes

Brown integral color with crushed red/brown 1/2 in. (12.5 mm) aggregate

Tan integral color with crushed gold/white/tan 1/2 in. (12.5 mm) aggregate

Uncolored concrete with round brown/gold/gray pebble

Buff/tan integral color with crushed green 1/2 in. (12.5 mm) aggregate

Dark gray integral color with crushed black/gray/white 1/2 in. (12.5 mm) aggregate

Brown integral color with 1/2 in. (12.5 mm) natural river pebble

Gold/tan integral color with round multi-colored 1/2 in. (12.5 mm) pebble

Gold/tan integral color with crushed gold/yellow 1/2 in. (12.5 mm) aggregate

Gold integral color with crushed yellow 1/2 in. (12.5 mm) aggregate

Tan/peach integral color with crushed yellow/tan 1/2 in. (12.5 mm) aggregate

Light red integral color with crushed pink 1/2 in. (12.5 mm) aggregate

Buff integral color with a crushed gold/gray/brown 3/8 to 1/2 in. (9.5 to 12.5 mm) granite

Integral Color with Seeded Aggregates

Gold/tan integral color with feldspar aggregate

Tan/peach integral color with feldspar aggregate

Medium gray integral color with feldspar aggregate

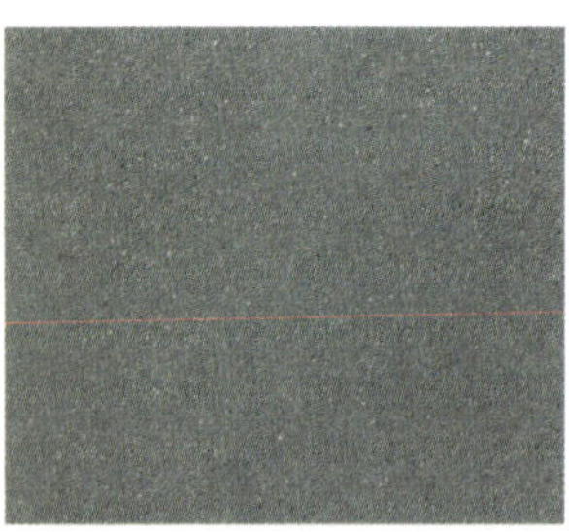

Dark gray integral color with silicon carbide aggregate

Sand Finishes Using Top Surface Retarder

All photos courtesy of EcoCast®

Uncolored concrete with a light sand finish using surface retarder

Brown integral color with a light sand finish using surface retarder

Light gray integral color with a light sand finish using surface retarder

Gold/tan integral color with a light sand finish using surface retarder

Red integral color with a light sand finish using top surface retarder

Medium gray integral color with a light sand finish using top surface retarder

Slate

Gold/tan integral color, textured with a slate texture mat in conjunction with a top surface retarder to expose sand

Light mauve integral color, textured with a slate texture mat in conjunction with a top surface retarder to expose sand

Stamped Concrete with Color Hardener and Contrasting Release Agent

Small sandstone stamped concrete using off-white color hardener with a medium brown release agent. Joints grouted with uncolored mortar

Ashlar slate stamped concrete using blue/gray color hardener with a charcoal release agent

12 x 12 in. (300 x 300 mm) slate stamped concrete using a beige color hardener with a medium brown release agent

Slate texture using a gray/green color hardener with a charcoal release agent

CHAPTER 12: PROJECT SAMPLES

Coastal Resort and Spa

Integral color pool deck using a light beige integral color with a sand finish achieved through the use of a top surface retarder. Concrete saw-cut at 4 ft (1.2 m) on center each way to minimize shrinkage cracking

Uncolored concrete patio with a sand finish. Concrete saw-cut 36 in. (0.9 m) on center each way

Cast-in-place uncolored concrete steps and walls. Walls have smooth finish and the steps have a sand finish as a result of a top surface retarder. The warning treads have been stained a dark brown to obtain code compliance

Civic Center Library and City Hall

Integral color plaza paving a light dose of integral color with a stamped slate texture in combination with a sand finish achieved through the use of a top surface retarder. Concrete saw-cut at 5 ft (1.5 m) on center each way to minimize shrinkage cracking and add interest to the design

Close-up of the colored, slate textured paving with a sand finish. The sand finish adds a natural touch to the stamped concrete

Colored concrete paving as it approaches the concrete stairs

Decorative Concrete Parking Lot

Integral color vehicular parking area with integral color sand finish paving and seeded aggregate banding. Both the sand finish and the seeded exposed aggregate banding were achieved through the use of top surface retarders. Surface retarders are available in varying degrees of etch to obtain desired results. Concrete saw-cut at 45 degrees achieves a visual contrast in the large paved area

Colored concrete paving surrounds the stone centerpiece in front of the office entry doors

Close-up of the colored, sand finish paving with exposed aggregate bands to designate the parking stalls

Hospital/Medical Center

Vehicular stamped concrete paving using an ashlar slate pattern and a light gray color hardener in combination with a contrasting charcoal release agent. The banding is comprised of an off-white color hardener with a medium gray release agent

Pedestrian concrete paving with an ashlar slate pattern in an off-white color hardener with a medium gray release agent. The banding is a light gray color hardener with a charcoal colored release agent

Church Plaza

Color hardeners in various colors with a sand finish. Note the vivid, bright, and durable colors that are achieved with color hardener. The contrast in the colors yields the compass design

Integral color and color hardener to achieve the design intent of the landscape architect

Close up of the colored concrete stepping stones with the adjacent concrete paving with a sand finish

Office Building Motor Court

Vehicular concrete paving using a dark gray integral color seeded with silicon carbide to yield a sparkle. In addition, the dark gray color tends to hide the marks left by tires at this vehicular turnaround. The banding is uncolored concrete seeded with silicon carbide. Both the vehicular paving and banding have a sand finish

Close-up of the dark gray silicon carbide seeded vehicular paving and uncolored silicon carbide seeded banding

Uncolored pedestrian paving seeded with silicon carbide and saw-cut at 24 in. (600 mm) on center each way

Regional Aquatics Center

Pedestrian concrete paving with a tan integral color and the use of a top surface retarder to achieve a consistent sand texture. The surface is nonslip and reduces glare

Close-up of the tan sand finish pedestrian paving. The concrete is saw-cut at 5 ft (1.5 m) on center each way to minimize shrinkage cracking

Medical Research Building

Pedestrian concrete paving with a white concrete and dark gray concrete to produce a symbolized DNA at this medical research campus. The white concrete was achieved by using white color hardener with a sand finish. White color hardener is durable and economical. The dark gray integral color banding has a sand finish achieved through the use of a top surface retarder

Close-up of the black and white pedestrian paving. The series of 6 in. (150 mm) wide bands were saw-cut. In addition, the banding is saw-cut transversely at 48 in. (1.2 m) on center to minimize cracking

Office Building

Integral color with a sand finish achieved by using a top surface retarder at a corporate headquarters

Overall view of the colored hardscape

Precast concrete truncated domes placed at a handicapped access ramp

Close-up detail showing isolation joints with caulking at the perimeter of a band. Note the placement of the saw-cut contraction joints to minimize shrinkage cracking

APPENDIX A

Curing Decorative Concrete

Problems and Practice #4

ACI-302.1R, "Guide for Concrete Floor and Slab Construction," covers concrete curing, but contains little information on curing decorative concrete. Some curing methods that are suitable for normal concrete may cause problems when used on decorative concrete and, in some cases, initial curing may have to be delayed. The following considerations should be addressed in a meeting with the architect, engineer, owner, and decorative concrete contractor during the planning stages of a decorative concrete project.

CURING CONSIDERATIONS:

- Traditional curing methods, especially water curing, may cause excessive discoloration and efflorescence.
- Monomolecular films or evaporation retarders may help prevent rapid moisture loss from the surface when the concrete is still in a plastic state.
- Slabs treated with liquid membrane-forming compounds will not be able to receive an acid stain treatment or an overlayment product until the compound is removed.
- Curing paper may be better suited for interior spaces scheduled for an acid-stain treatment.
- Don't rely on dissipating liquid membrane forming compounds to make a surface suitable for acid staining or overlays because the time required to dissipate can vary and is affected by the amount of sunlight exposure.
- Slabs treated with a powdered release agent and textured with mats are unable to accept liquid membrane-forming compounds until the release agent is washed off one to three days after the job is completed.
- Combination sealers that both cure and seal have been created that meet ASTM 1315 curing standards while providing a decorative finish. They can be applied as soon as release has been removed and the surface is dry, many times within a 24-hour window. Manufacturers' recommendations should always be followed for the specific product and or application.
- Colored concrete placements without texture have the option of colored or tinted curing agents that will help level the color over multiple pours. Time must be allowed for the curing agent to dissipate prior to the application of any traditional decorative sealer.

A different curing approach may be needed for different types of decorative concrete applications. Before the project starts, it is imperative for the construction team to be aware of the results a curing method will produce. The specified curing method should always be applied to the mock-up for the project prior to the start of the project. The mock-up should be treated with the same coloring, texturing and sealing methods to be used on the actual project. The decorative concrete contractor must be involved in this decision to ensure that the project will meet the specified requirements of the owner.

Decorative concrete contractors will work with owners and architects in developing specifications and constructing mock-ups to address their appearance requirements. If you have any questions, contact your ASCC/DCC concrete contractor or the ASCC Decorative Concrete Hotline at (888) 483-5288.

ASCC Decorative Concrete Council
A Specialty Council of the American Society of Concrete Contractors
2025 S. Brentwood Blvd. Suite 105 ■ St. Louis, MO 63144
Telephone: 314-962-0210 ■ Website: www.ascconline.org
Toll Free: 866-788-2722 ■ E-mail: ascc@ascconline.org

Get Certified
Since 1980, ACI has tested over 400,000 concrete technicians, inspectors, supervisors, and craftsmen in 20 different certification programs.
When you have a need for qualified concrete professionals—specify ACI Certification.
Visit www.ACICertification.org for:
Descriptions of ACI Certification Programs — Includes program requirements and reference/resource materials.
Schedule of Upcoming/Testing Sessions — Search by program and/or state.
Directory of Certified Individuals — Confirm an individual's certification and date of expiration.

9 781942 727590